MATHEMATICAL SURVEYS AND MONOGRAPHS SERIES LIST

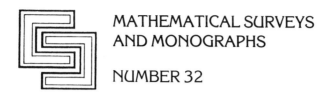

MATHEMATICAL SURVEYS
AND MONOGRAPHS

NUMBER 32

AN INTRODUCTION
TO CR STRUCTURES

HOWARD JACOBOWITZ

American Mathematical Society
Providence, Rhode Island

1980 *Mathematics Subject Classification* (1985 *Revision*). Primary 32F25; Secondary 32-02, 53-02.

Library of Congress Cataloging-in-Publication Data

Jacobowitz, Howard, 1944-
    An introduction to CR structures/Howard Jacobowitz.
        p. cm. -- (Mathematical surveys and monographs; no. 32)
    Includes bibliographical references (p.) and index.
    ISBN 0-8218-1533-4 (alk. paper)
    1. CR submanifolds. 2. Geometry, Differential. I. Title. II. Series.
QA649.J33   1990   90-608
516.3'6--dc20   CIP

To my parents

Contents

Preface

In 1907 Poincaré showed that two real hypersurfaces in $\mathbb{C}^2$ are in general biholomorphically inequivalent, [**Po**]. That is, given two real analytic submanifolds of real dimension three in $\mathbb{C}^2$ there usually is no biholomorphism of one open set in $\mathbb{C}^2$ to another open set in $\mathbb{C}^2$ that takes a piece of the first submanifold onto the second. He then raised the question of finding the invariants that distinguish one real hypersurface from another. This basic question was first completely answered by Cartan, in [**Ca2**]. Cartan remarked "Je reprends la question directement comme application de ma méthode générale d'équivalence. La résolution complète du problème de Poincaré me conduit à des notions géométriques nouvelles ... ". A second solution was given by Moser in 1973. In joint work with Chern [**CM**] this was generalized, along with Cartan's original solution, to dimensions greater than two. (At about this time, Tanaka in [**Tan1**] and [**Tan2**] gave a different extension of Cartan's work to higher dimensions.)

The study of the basic problem for CR structures primarily rests on these two works. Thus, although most introductions to an area of mathematics are a synthesis from many sources, reflecting how mathematics usually develops, the present one, to a surprising degree, is not. Rather, it is in large measure an exposition of the papers of Cartan and of the joint paper of Chern and Moser.

For Cartan one needs to know something of his general method of equivalences and of the structure of the Lie group $SU(2,1)$. So we have tried to provide this necessary background before going over the main construction. This background is also important for the part of the Chern-Moser paper that extends Cartan's work to higher dimension. For the rest of the Chern-Moser paper the problem is somewhat different. Here the approach is more straightforward but also more technically difficult. We do the lowest-dimensional case in detail. This should also make the higher-dimensional case more accessible.

An exception to this focus on the above two works is Chapter 1. Here we give the basic definitions and properties and draw from many sources in the literature and in the "folklore." Chapter 2 uses simple facts about the automorphism groups of the ball and polydisc in order to show that the Riemann Mapping Theorem does not hold in higher dimension. Then the automorphism group of the ball, $SU(2, 1)$, is computed in enough detail for future needs. The next two chapters go over the Moser normal form. In Chapter 5, we give the background necessary to understand Cartan's work. Then in Chapter 6 we present Cartan's basic construction along with two variations, one of which is the lowest-dimensional case of the construction in the Chern-Moser paper. In the next three chapters we explore some aspects of Cartan's "new geometric ideas," and also relate his invariants to those found by Moser. Finally, in Chapter 10 we consider the landmark paper of Hans Lewy [**Le2**] and the realizability problem for abstract CR structures. Thus we end with a certain historical completeness — Professor Lewy has stated that his interest in the partial differential operators associated to real hypersurfaces arose while trying to understand the papers of Cartan.

A number of exercises are included; some to help the reader clarify material for himself or herself and others to simplify the author's task. After Chapter 10 there are Notes providing additional information for some of the chapters and also indicating some of the material that would have been included if the author had not been anxious to conclude this long delayed project.

This book evolved from lecture notes for courses given at Rutgers University and Université de Grenoble. The author is grateful for these opportunities to lecture on this subject. The author is also grateful to the National Science Foundation for its support of his research, some of which has found its way into this book, and to Provost Walter Gordon of Rutgers-Camden for his support of this project. Finally, it is a pleasure to acknowledge the constant encouragement and support of Dalia Ritter.

CHAPTER 1

CR Structures

§1. Some observations of Poincaré. This book studies the induced structure on real hypersurfaces M^{2n+1} in the complex space $\mathbb{C}^{n+1}$ and a generalization to abstract manifolds. Most of the phenomena and difficulties already occur for $n = 1$; we generally restrict ourselves to this case. The subject arose from a paper of Poincaré in which he considered whether the Riemann Mapping Theorem could be generalized from $\mathbb{C}^1$ to $\mathbb{C}^2$. He gave two simple arguments that even a local version, which is trivial in $\mathbb{C}^1$, does not hold in higher dimensions and he also laid the basis for a global counterexample. This chapter starts with his local arguments; the next with his global one.

Any two real analytic curves in $\mathbb{C}^1$ are locally equivalent: Given points p and q on the curves Γ_1 and Γ_2 there are open subsets of $\mathbb{C}^1$, U_1 containing p and U_2 containing q, and a biholomorphism $\Phi: U_1 \to U_2$ with $\Phi(U_1 \cap \Gamma_1) = U_2 \cap \Gamma_2$. This may be taken as a very weak form of the Riemann Mapping Theorem. Poincaré showed that the analogous result does not hold in $\mathbb{C}^2$. Namely, let s and S be real analytic surfaces of real dimension three in $\mathbb{C}^2$. In general, there will not be a local biholomorphism taking one to the other. The first of Poincaré's two proofs of this uses the fact that a function on a hypersurface is the restriction of a holomorphic function only if it satisfies a certain partial differential equation. We give Poincaré's derivation of this fact (using mostly his notation). Let the hypersurface be written as a graph

$$s = \{(x_1 + iy_1, x_2 + iy_2): x_1 = \phi(y_1, x_2, y_2)\}$$

and let $F(y_1, x_2, y_2)$ be the function on s. We ask: Does there exist a holomorphic function $f(z_1, z_2)$ of $z_1 = x_1 + iy_1$ and $z_2 = x_2 + iy_2$ such that $F(y_1, x_2, y_2) = f(\phi(y_1, x_2, y_2) + iy_1, x_2 + iy_2)$? If so

$$\frac{\partial F}{\partial y_1} = \frac{\partial f}{\partial z_1}(\phi_{y_1} + i) \quad \text{and} \quad \frac{\partial F}{\partial \bar{z}_2} = \frac{\partial f}{\partial z_1} \phi_{\bar{z}_2}.$$

If this notation is not familiar, look at the next section where it is formally introduced. Let

$$(1) \qquad\qquad L = \phi_{\bar{z}_2} \frac{\partial}{\partial y_1} - (i + \phi_{y_1}) \frac{\partial}{\partial \bar{z}_2}$$

Thus $LF = 0$.

EXERCISE 1. Compute L for the hypersurfaces

$$x_1 = x_2^2 + y_2^2$$

and

$$x_1 = 0.$$

EXERCISE 2. Let Ψ be a biholomorphism and s a hypersurface. Let $S = \Psi(s)$, write S as a graph, and compute its operator L_S. Show that $\Psi_*(L) = \lambda L_S$ where λ is a nonzero complex function. This may be interpreted as meaning that although L itself depends on the choice of local coordinates for $\mathbb{C}^2$ the set of multiples $\{\lambda L\}$ depends only on the hypersurface s.

We consider two surfaces

$$s = \{x_1 = \phi(y_1, x_2, y_2)\},$$
$$S = \{X_1 = \Phi(Y_1, X_2, Y_2)\}.$$

For s and S to be locally equivalent we need to find three real functions Y_1, X_2, Y_2 of the variables (y_1, x_2, y_2) such that both $f_1 = \phi(Y_1, X_2, Y_2) + iY_1$ and $f_2 = X_2 + iY_2$ are the restrictions of holomorphic functions. That is, we must solve for $j = 1$ and 2

$$\left(\phi_{\bar{z}_2}(y_1, x_2, y_2) \frac{\partial}{\partial y_1} - (i + \phi_{y_1}(y_1, x_1, y_2)) \frac{\partial}{\partial \bar{z}_2} \right) f_j = 0.$$

We have four real equations for three real unknowns. Thus solutions usually do not exist.

EXERCISE 3. Show that $x_1 = x_2^2 + y_2^2$ and $x_1 = 0$ are not locally equivalent by making this argument rigorous for these two surfaces.

Poincaré's second demonstration that in general hypersurfaces are inequivalent is also a counting argument (and contrasts with the previous one in the same way that the power series definition of holomorphic function contrasts with the differential equation definition). We start with a simple observation and an elegant proof (which the author learned from E. Calabi.)

LEMMA 1. *The Taylor series expansion up to and including terms of order N of a function of k variables contains $\binom{N+k}{k}$ terms.*

PROOF. The number we seek is clearly also the number of homogeneous monomials of order N in $k + 1$ variables. We may write each monomial uniquely as $x_{i_1} x_{i_2} \cdots x_{i_N}$ with $i_1 \leq i_2 \leq \cdots \leq i_N$. Thus we need to know the number of elements in the set

$$I = \{(i_1, \ldots, i_N) \mid 1 \leq i_1 \leq i_2 \leq \cdots \leq i_N \leq k + 1\}.$$

Let

$$J = \{(j_1, \ldots, j_N) \mid 2 < j_1 < j_2 < \cdots < j_N < N + k + 1\}.$$

There are $\binom{N+k}{N}$ elements in J. For each $\{i_p\} \in I$, consider the new sequence $\phi\{i_p\} = \{i_1 + 1, i_2 + 2, \ldots, i_N + N\}$. Note that $\phi\{i_p\}$ is in J and that ϕ is both injective and surjective. Thus there are also $\binom{N+k}{N}$ elements in I. Since $\binom{N+k}{N} = \binom{N+k}{k}$, this proves the lemma.

Let us fix some hypersurface M given by $v = F_0(x, y, u)$, containing the origin. Then any nearby surface is of the same form $v = F(x, y, u)$. There is a map from these hypersurfaces into $\mathbb{R}^p$, $p = \binom{N+3}{3}$, given by the N-jet at the origin of F. This map clearly covers a neighborhood of the origin. Now let us consider the set $\mathscr{S}$ of local biholomorphisms of $\mathbb{C}^2$ defined in a neighborhood of the origin and taking M to a hypersurface of the same form. Thus we also have a map of $\mathscr{S}$ into $\mathbb{R}^p$. This map factors through the map of $\mathscr{S}$ into the space of N-jets of pairs of holomorphic functions, namely $\mathbb{R}^q$, $q = 4\binom{N+2}{2}$. If M were equivalent to every other hypersurface of the same form there would be a continuous map from $\mathbb{R}^q$ to $\mathbb{R}^p$ which covers some neighborhood of the origin. But this is impossible whenever $N > 10$ because in this case $p > q$. This is our second proof that in general hypersurfaces are inequivalent.

Now assume that points p_1 and p_2 are given on hypersurfaces M_1 and M_2 and we ask what is the highest order of contact possible, in general, between M_2 and $\Phi(M_1)$ where Φ is a biholomorphism taking p_1 to p_2. We may assume $p_1 = p_2 = 0$ and M_1 and M_2 both have the form $v = F(z, y, u)$. Then to make the N-jet of $\Phi(M_1)$ agree with the N-jet of M_2 involves $\binom{N+3}{3} - 1$ conditions, while the N-jet of Φ involves only $4\{\binom{N+2}{2} - 1\}$ parameters. Thus, as Poincaré pointed out, one cannot achieve ninth-order contact. In fact, as we shall see in Chapter 3, an obstruction already is present to sixth-order contact.

These two simple arguments already point to our major topics. Since not all hypersurfaces are locally equivalent it is natural to seek invariants which allow us to distinguish one from another. The problem of finding these invariants in $\mathbb{C}^2$ was solved by Cartan [Ca2] and in a completely different manner by Moser and then generalized to higher dimensions by Chern and Moser [CM]. These two solutions to the problem posed by Poincaré will occupy us for most of this book. And the remainder can be traced to the observation that a function on a hypersurface is the restriction of a holomorphic function only if it satisfies a certain differential equation.

§**2. CR manifolds.** We start with some standard notation and concepts. Let $(z_1, \ldots, z_n)$ be the usual coordinates for $\mathbb{C}^n$ and $(x_1, y_1, \ldots, x_n, y_n)$ the corresponding coordinates for $\mathbb{R}^{2n}$. We define the first-order partial differential operation

$$\frac{\partial}{\partial z_j} = \frac{1}{2}\left(\frac{\partial}{\partial x_j} - i\frac{\partial}{\partial y_j}\right)$$

and its conjugate operation

$$\frac{\partial}{\partial \bar{z}_j} = \frac{1}{2}\left(\frac{\partial}{\partial x_j} + i\frac{\partial}{\partial y_j}\right).$$

The conventions are easily explained: A function $F(z) = f(x, y)$ of one complex variable is holomorphic if and only if $\frac{\partial}{\partial \bar{z}} f = 0$ and the factor of $\frac{1}{2}$ is necessary for $dz \left(\frac{\partial}{\partial z} \right) = 1$ where $dz = dx + i\, dy$.

We may also consider $\frac{\partial}{\partial z_j}$ and $\frac{\partial}{\partial \bar{z}_j}$ as complex vector fields. If V is a vector space over the real numbers then $\mathbb{C} \otimes V$ is the corresponding vector space over the complex numbers. If $\{v_1, \ldots, v_n\}$ is a basis for V, then

$$\mathbb{C} \otimes V = \left\{ \sum_{j=i}^{N} \alpha_j v_j \; : \; \alpha_j \in \mathbb{C} \right\}.$$

Next consider a manifold M and the tangent space $T_p M$ to M at a point p. The *tangent bundle* is given by

$$TM = \bigcup_p T_p M$$

and the *complexified tangent bundle* by

$$\mathbb{C} \otimes TM = \bigcup_p \mathbb{C} \otimes T_p M.$$

When M is $\mathbb{R}^{2n}$, we have as a basis for TM

$$\left\{ \frac{\partial}{\partial x_1}, \frac{\partial}{\partial y_1}, \ldots, \frac{\partial}{\partial x_n}, \frac{\partial}{\partial y_n} \right\}$$

and as a basis for $\mathbb{C} \otimes TM$

$$\left\{ \frac{\partial}{\partial z_1}, \ldots, \frac{\partial}{\partial z_n}, \frac{\partial}{\partial \bar{z}_1}, \ldots, \frac{\partial}{\partial \bar{z}_n} \right\}.$$

We have seen that there is a differential operator L defined on $M^3 \subset \mathbb{C}^2$ with the property that $Lf = 0$ if f is the restriction to M of a holomorphic function. So L may be considered as the induced Cauchy–Riemann operator. We shall soon see that for $M^{2n+1} \subset \mathbb{C}^{n+1}$ there are n such induced Cauchy–Riemann operators. This is what is abstracted as the definition of CR manifolds. To explain this, it is best to start with the almost complex structure of $\mathbb{C}^{n+1}$.

DEFINITION. Let M be a manifold and let V be a sub-bundle of $\mathbb{C} \otimes TM$. Then (M, V) is an *almost complex manifold* if

$$V \cap \bar{V} = \{0\}$$

and

$$V \oplus \bar{V} = \mathbb{C} \otimes TM.$$

Set $\dim_{\mathbb{C}} V = n$; it follows that $\dim_{\mathbb{R}} M = 2n$.

If M is a complex manifold then the underlying almost complex manifold is given locally by choosing complex coordinates and setting

(2) $$V = \text{linear span}_{\mathbb{C}} \left\{ \frac{\partial}{\partial \bar{z}_1}, \ldots, \frac{\partial}{\partial \bar{z}_n} \right\}.$$

This sub-bundle is clearly independent of the choice of coordinates since it is preserved by a holomorphic transformation. We use $L \in V$ as an abbreviation for "L is a section of V over an appropriate set." Note that a function f is holomorphic if $Lf = 0$ for all $L \in V$. So V is the bundle of Cauchy–Riemann operators.

Here is another definition of an almost complex manifold.

DEFINITION. Let M be a manifold and let $J: TM \to TM$ be a fiber isomorphism. Then (M, J) is an *almost complex manifold* if

$$J^2 = -\operatorname{Id}.$$

If M is a complex manifold then the underlying almost complex manifold is given by

(3) $$J\left(\frac{\partial}{\partial \overline{z}_k}\right) = -i\left(\frac{\partial}{\partial \overline{z}_k}\right) \quad \text{and} \quad J\left(\frac{\partial}{\partial z_k}\right) = i\left(\frac{\partial}{\partial z_k}\right).$$

Again, J is independent of the choice of coordinates. For if

$$z_k = \phi_k(w_1, \ldots, w_n)$$

is a change of coordinates, and if J is defined using z, then

$$J\left(\frac{\partial}{\partial w_k}\right) = J\left(\frac{\partial \phi_j}{\partial w_k}\frac{\partial}{\partial z_j}\right) = \frac{\partial \phi_j}{\partial w_k}\left(i\frac{\partial}{\partial z_j}\right) = i\frac{\partial}{\partial w_k}$$

and in the same way

$$J\left(\frac{\partial}{\partial \overline{w}_k}\right) = -i\frac{\partial}{\partial \overline{w}_k}.$$

So J coincides with the operator defined using w.

Note that J is real in the sense that it provides a map of TM to itself, since, if X is a real tangent vector then

$$X = \sum\left(\alpha_j\frac{\partial}{\partial z_j} + \overline{\alpha}_j\frac{\partial}{\partial \overline{z}_j}\right)$$

and

$$JX = \sum\left(i\alpha_j\frac{\partial}{\partial z_j} - \overline{\alpha}_j\frac{\partial}{\partial \overline{z}_j}\right).$$

So JX is also a real tangent vector.

It is easy to see that these two definitions of almost complex manifolds are equivalent. For given (M, V), we define $J: \mathbb{C} \otimes TM \to \mathbb{C} \otimes TM$ by letting V and $\overline{V}$ be its eigenspaces corresponding to the eigenvalues $-i$ and i respectively. It is clear that such a J restricts to a map of TM to itself and satisfies $J^2 = -\operatorname{Id}$. Conversely, given $J: TM \to TM$ with $J^2 = -\operatorname{Id}$, we extend J linearly to a map of $\mathbb{C} \otimes TM$ to itself and let V be the eigenspace corresponding to the eigenvalue $-i$.

Let X be any real tangent vector. Note that

(4) $$J(X + iJX) = -i(X + iJX).$$

Thus

(5) $V = \{X + iJX : X \in TM\}.$

REMARK. The J-operator on $\mathbb{R}^2$ is simply counterclockwise rotation by $90°$ and the J-operator on $\mathbb{C}$ is multiplication by i.

EXERCISE 4.

(a) Show that a function $f : M \to \mathbb{C}$ on a complex manifold is holomorphic if and only if $df \cdot J_1 = J_2 \cdot df$ where J_1 gives the complex structure on M and J_2 gives the complex structure on $\mathbb{C}$.

(b) A function f on a complex manifold is called *anti-holomorphic* if $\frac{\partial}{\partial z_j} f(z, \bar{z}) = 0$ for each coordinate z_j. Show that f is anti-holomorphic if and only if

$$df \cdot J_1 = -J_2 \cdot df.$$

Not every almost complex manifold comes from some complex manifold. It is easy to see that a necessary condition for an almost complex manifold to be complex is that

(6) $[V, V] \subset V.$

By this we mean that the commutator (or bracket) of any two local sections of V is still a local section of V. This is clear using local coordinates

$$\left[a_j \frac{\partial}{\partial \bar{z}_j}, b_k \frac{\partial}{\partial \bar{z}_k} \right] = \left(a_j \frac{\partial b_k}{\partial \bar{z}_j} - b_j \frac{\partial a_k}{\partial \bar{z}_j} \right) \frac{\partial}{\partial \bar{z}_k}.$$

Using (3) we see that in terms of J this means

(7) $[X, JY] + [JX, Y] = J\{[X, Y] - [JX, JY].\}$

The two equivalent conditions (6) and (7) are necessary for an almost complex manifold to be complex. An almost complex manifold which satisfies these conditions is said to be integrable. There is an easy result and a hard result about such manifolds. It is no loss of generality to assume that M is a real analytic manifold. Fix some coordinate system from this real analytic atlas. If in this coordinate system V has a basis of vectors with real analytic complex-valued coefficients, or, equivalently, if J is given by a real analytic matrix, then the almost complex structure is said to be real analytic.

THEOREM (EASY). *A real analytic, integrable almost complex structure is complex.*

NEWLANDER–NIRENBERG THEOREM. [NN] *A* $C^{1+\lambda}$ *integrable almost complex structure is complex.*

In terms of the usual coordinates on $\mathbb{R}^{2n}$, the J operator in (3) is

$$J\left(\frac{\partial}{\partial x_k} \right) = \frac{\partial}{\partial y_k}, \quad J\left(\frac{\partial}{\partial y_k} \right) = -\frac{\partial}{\partial x_k}.$$

The tangent space to $\mathbb{R}^{2n}$ at a point can be identified with $\mathbb{R}^{2n}$ itself. The map $J: \mathbb{R}^{2n} \to \mathbb{R}^{2n}$ is thus given by

$$J(x_1, y_1, \ldots, x_n, y_n) = (-y_1, x_1, \ldots, -y_n, x_n).$$

The corresponding map $J: \mathbb{C}^n \to \mathbb{C}^n$ is

$$J(z_1, \ldots, z_n) = (iz_1, \ldots, iz_n).$$

Here is a notation which is useful as our focus switches between $\mathbb{R}^{2n}$ and $\mathbb{C}^n$. For $x = (x_1, y_1, x_2, y_2, \ldots, x_n, y_n) \in \mathbb{R}^{2n}$, let $X_{\mathbb{C}} = (x_1 + iy_1, x_2 + iy_2, \ldots, x_n + iy_n) \in \mathbb{C}^n$ and for $Z = (x_1 + iy_1, x_2 + iy_2, \ldots, x_n + iy_n) \in \mathbb{C}^n$ let $Z_{\mathbb{R}} = (x_1, y_1, x_2, y_2, \ldots, x_n, y_n)$. So

$$JX = (iX_{\mathbb{C}})_{\mathbb{R}}$$

and

$$(JX)_{\mathbb{C}} = iX_{\mathbb{C}}.$$

Later, we will need to find a vector U orthogonal to both X and JX. We use the notation

$$Z \cdot W = \sum_1^n Z_j W_j$$

for vectors in $\mathbb{C}^n$ and

$$X \cdot U = \sum_1^{2n} X_j U_j$$

for vectors in $\mathbb{R}^{2n}$. Then for $Z = X_{\mathbb{C}}$ and $W = U_{\mathbb{C}}$

$$W \cdot \overline{Z} = U \cdot X + iU \cdot JX.$$

So U is orthogonal to both X and JX if and only if

(8) $$U_{\mathbb{C}} \cdot \overline{X}_{\mathbb{C}} = 0.$$

LEMMA 2. *Let* $X_1, \ldots, X_p$ *be the vectors in* $\mathbb{R}^{2n}$. *The set* $\{X_1, \ldots, X_p, JX_1, \ldots, JX_p\}$ *is linearly independent over* $\mathbb{R}$ *if and only if the set* $\{X_1 - iJX_1, \ldots, X_p - iJX_p\}$ *is independent over* $\mathbb{C}$.

PROOF. For any real vector Y and scalars a and b

$$(a + ib)(Y - iJY) = (aY + bJY) - iJ(aY + bJY);$$

so

$$\sum_1^p (a_j + ib_j)(X_j - iJX_j) = \sum_1^p (a_j X_j + b_j JX_j) - iJ\left(\sum_1^p a_j X_j + b_j JX_j\right).$$

The rest is obvious.

We now define a totally real subspace of $\mathbb{R}^{2n}$ (or equivalently of $\mathbb{C}^n$) and then use this to define totally real submanifolds of complex manifolds.

We think of $\mathbb{R}^{2n}$ as a vector space and identify linear subspaces with planes through the origin.

DEFINITION. A linear subspace P of $\mathbb{R}^{2n}$ is *totally real* if it has a basis $\{X_1, X_2, \ldots, X_p\}$ such that the set $\{X_1, X_2, \ldots, X_p, JX_1, JX_2, \ldots, JX_p\}$ is linearly independent.

It is easy to see that if one basis satisfies this then so does every other basis. In other words, P is totally real if any basis for P remains linearly independent when J of each vector is added to the basis.

DEFINITION. A real linear subspace Q of $\mathbb{C}^n$ is *totally real* if it has a basis over $\mathbb{R}$, $\{Z_1, \ldots, Z_q\}$, such that the set $\{Z_1, \ldots, Z_q, iZ_1, \ldots, iZ_q\}$ is linearly independent over $\mathbb{R}$.

Again, this property is independent of the particular basis. Q is totally real if any basis for Q is also linearly independent over $\mathbb{C}$.

That these two definitions are consistent with the identification of $\mathbb{C}^n$ with $\mathbb{R}^{2n}$ is easily seen. For if $Z_k = (X_k)_\mathbb{C}$ then

$$\sum(a_k X_k + b_k J X_k) = \sum((a_k + ib_k)Z_k)_\mathbb{R}.$$

LEMMA 3.
(a) P is totally real if and only if $X \in P$ implies $JX \notin P$.
(b) Q is totally real if and only if $Z \in Q$ implies $iZ \notin Q$.

EXAMPLE. The real 2-plane in $\mathbb{C}^2$ spanned by the vectors $(1, 0)$ and $(i, 0)$ is not totally real. The real 2-plane spanned by $(1, 0)$ and $(0, i)$ is totally real and coincides with the plane in $\mathbb{R}^4$ spanned by $(1, 0, 0, 0)$ and $(0, 0, 0, 1)$.

Here is a simple exercise that will be used in Chapter 9. Let A be an $n \times n$ complex matrix. Call a vector X in $\mathbb{C}^n$ a *conjugate fixed point* of A if $A\overline{X} = X$.

EXERCISE 5. The set of conjugate fixed points is a totally real subspace of $\mathbb{C}^n$.

DEFINITION. A submanifold S of the complex manifold M is *totally real* if at each point $p \in S$ the tangent space $T_p S$ is a totally real subspace of $T_p M$.

REMARKS.
1. An equivalent condition is

$$T_p S \cap J(T_p S) = \{0\}.$$

2. A related concept is for S to be *maximally real*. This means that S is totally real and of the highest dimension, namely, $\dim S = n$.

We now consider real hypersurfaces of a complex space. Let M^{2n+1} be a submanifold of $\mathbb{C}^{n+1}$. We show how each of our definitions of an almost complex structure on $\mathbb{C}^{n+1}$ leads to an induced structure on M and then that these two induced structures are essentially the same. So first we set

$$H_{0,1}(\mathbb{C}^{n+1}) = \text{linear span}_{\mathbb{C}} \left\{ \frac{\partial}{\partial \bar{z}_1}, \dots, \frac{\partial}{\partial \bar{z}_{n+1}} \right\} \subset \mathbb{C} \otimes T\mathbb{C}^{n+1}$$

and define $H_{0,1}(M)$ by

$$H_{0,1}(M) = H_{0,1}(\mathbb{C}^{n+1}) \cap \mathbb{C} \otimes TM.$$

When no confusion is possible we denote $H_{0,1}(M)$ by V. So

$$V \cap \bar{V} = \{0\}.$$

Note that $w \in V$ if

$$w = \sum \alpha_k \frac{\partial}{\partial \bar{z}_k}$$

and in addition w is "tangent" to M.

DEFINITION. A complex vector W is *tangent* to M if the real vectors $\operatorname{Re} W$ and $\operatorname{Im} W$ are tangent to M.

It is clear that W is tangent to M if and only if $Wr = 0$ for any choice of a defining function for M. Note that we have

(9) $$V \cap \bar{V} = \{0\} \quad \text{and} \quad \dim_{\mathbb{C}} V = n.$$

We also could use the second definition of an almost complex manifold to define the structure induced on $M^{2n+1} \subset \mathbb{C}^{n+1}$. To do this we let $H = TM \cap JTM$ and consider the restriction to H of J. We have

(10) $$H \subset TM, \quad \dim_{\mathbb{R}} H = 2n, \quad J: H \to H, \quad J^2 = -\operatorname{Id}.$$

Since the almost complex structure on $\mathbb{C}^{n+1}$ is complex, we have that the necessary conditions (6) and (7) hold. Thus the induced structure on M defined by $H_{0,1}(M)$ satisfies

(11) $$[V, V] \subset V.$$

Note that a vector W is in H if both W and JW are tangent to M. Thus from (7) we see that the structure on M defined using $H = TM \cap JTM$ satisfies for X and Y in H

$$[JX, Y] + [X, JY] \in H$$

(12) and

$$J\{[JX, Y] + [X, JY]\} = [JX, JY] - [X, Y].$$

EXERCISE 6. Show that if M_1 and M_2 are real hypersurfaces in $\mathbb{C}^{n+1}$ and if $\Phi: U_1 \to U_2$ is a biholomorphism of open subsets of $\mathbb{C}^{n+1}$ with $\Phi(U_1 \cap M_1) = U_2 \cap M_2$ then $\Phi_* V_1 = V_2$ and $\Phi_* H_1 = H_2$.

We are now ready to define the basic object of study in these notes. We start with an abstract odd-dimensional manifold M.

DEFINITION. (M, V) is a *CR manifold* if $\dim_{\mathbb{R}} M = 2n + 1$, V is a sub-bundle of $\mathbb{C} \otimes TM$ with $\dim_{\mathbb{C}} V = n$, $V \cap \bar{V} = \{0\}$, and $[V, V] \subset V$.

DEFINITION. (M, H, J) is a *CR manifold* if $\dim_{\mathbb{R}} M = 2n + 1$, H is a sub-bundle of TM with $\dim_{\mathbb{R}} H = 2n$, $J : H \to H$, and $J^2 = -\text{Id}$.

If X and Y are in H, then so is $[JX, Y] + [X, JY]$ and $J\{[JX, Y] + [X, JY]\} = [JX, JY] - [X, Y]$.

REMARKS.

1. The CR refers to Cauchy–Riemann because for $M \subset \mathbb{C}^{n+1}$, V consists of the induced Cauchy–Riemann operators. That is, a function f on M can be the restriction of a holomorphic function on an open subset of $\mathbb{C}^{n+1}$ only if $Lf = 0$ for all sections L of V. Compare this to Poincaré's definition of (1).

2. These structures might more precisely be called CR manifolds of hypersurface type to reflect the fact that we generalized the structure of hypersurfaces in $\mathbb{C}^{n+1}$ rather than the structure of submanifolds of higher codimension.

3. Almost everything that we will be doing will be local in nature. So often we refer to (M, V) or (M, H, J) as a CR structure rather than a CR manifold to emphasize that our study is of local properties rather than of global objects in the manifold.

4. A real sub-bundle of TM is called an r-plane distribution, where r is the fiber dimension of the sub-bundle. So here H is a $2n$-plane distribution on M. The fiber of H at a point p is called the *characteristic plane* at p.

5. For M^3, there are no integrability conditions and (11) and (12) can be ignored: Any complex line bundle V with $V \cap \overline{V} = \{0\}$, or any 2-plane distribution H together with a J operator, defines a CR structure.

6. The induced structure on hypersurfaces in $\mathbb{C}^{n+1}$ led to the definition of an abstract CR manifold. It is natural to wonder if this concept is in fact more general than the motivating examples. That is, can every CR structure (M^{2n+1}, V) be realized, at least locally, by a hypersurface in $\mathbb{C}^{n+1}$? This is a fascinating question. We discuss it in Section 4 and come back to it in Chapter 10.

Let us show that the two definitions of a CR structure are equivalent. Given V we choose some basis $\{L_1, \ldots, L_n\}$ and note that $V \cap \overline{V} = \{0\}$ implies that $\{\text{Re } L_1, \ldots, \text{Im } L_n\}$ is linearly independent. Set H equal to the linear space over $\mathbb{R}$ of this set and define J by $J(\text{Re } L_k) = \text{Im } L_k$, $J(\text{Im } L_k) = -\text{Re } L_k$. H clearly does not depend on the choice of basis. The map J extends to a complex linear map of $\mathbb{C} \otimes H$ to itself with V as its $-i$ eigenspace and $\overline{V}$ as its $+i$ eigenspace. So J also is independent of the choice of basis. The integrability condition for J follows from that for V.

Given H and J, extend J to a complex linear map of $\mathbb{C} \otimes H$ to itself and let V be the $-i$ eigenspace. Clearly $V \cap \overline{V} = \{0\}$. And now the integrability condition for J implies the one for V.

In summary, the condition

$$L \in V$$

is equivalent to

$$L = X + iJX \quad \text{for some} \quad X \in H.$$

We show two ways of computing V for $M^3 \subset \mathbb{C}^2$. Since $\dim_{\mathbb{C}} V = 1$, we need only find a complex vector field L which is tangent to M and has the form

$$\alpha \frac{\partial}{\partial \overline{z}} + \beta \frac{\partial}{\partial \overline{w}}.$$

So let r be a defining function for M,

$$M = \{(z, \overline{z}, w, \overline{w}): r(z, \overline{z}, w, \overline{w}) = 0\}.$$

We need that $Lr = 0$ and so we can take

(13) $$L = r_{\overline{w}} \frac{\partial}{\partial \overline{z}} - r_{\overline{z}} \frac{\partial}{\partial \overline{w}}.$$

This relies on the extrinsic coordinates z and w.

We may also express L intrinsically. Let M be given as a graph,

$$M = \{(x, y, u, v): v = \phi(x, y, u)\}.$$

The operators $\frac{\partial}{\partial u}$ and

$$\frac{\partial}{\partial z} = \frac{1}{2} \left(\frac{\partial}{\partial x} - i \frac{\partial}{\partial y} \right),$$

$$\frac{\partial}{\partial \overline{z}} = \frac{1}{2} \left(\frac{\partial}{\partial x} + i \frac{\partial}{\partial y} \right)$$

are here thought of as intrinsic on M. Any complex vector field on M has the form

$$L = \alpha \frac{\partial}{\partial z} + \beta \frac{\partial}{\partial \overline{z}} + \gamma \frac{\partial}{\partial u}.$$

We want L to be the restriction of a Cauchy–Riemann operator on $\mathbb{C}^2$. This means that we should have $Lf = 0$ if $f = F(z, u + i\phi(x, y, u))$ where $F(z, w)$ is holomorphic. This leads to

$$\alpha = 0 \quad \text{and} \quad \beta i \phi_{\overline{z}} = -\gamma(1 + i\phi_u).$$

We are only interested in L up to multiples so we set $\beta = 1 + i\phi_u$. This gives

(14) $$L = (1 + i\phi_u) \frac{\partial}{\partial \overline{z}} - i\phi_{\overline{z}} \frac{\partial}{\partial u}.$$

Note that this is the same as the operator (1) introduced by Poincaré but written with respect to a different choice of coordinates. There is a useful trick which lets us verify that (13) and (14) are the same operators (up to multiple). Let $f(z, \overline{z}, u)$ be given. Since the operator in (13) is tangent to M we may compute Lf by extending f in an arbitrary way off of M. The value of Lf does not depend on the extension. If we take the extension to be independent of v, then $f_{\overline{w}} = \frac{1}{2}f_u$. Since

$$r = \frac{w - \overline{w}}{2i} - \phi(z, \overline{z}, u)$$

we have that

$$Lf = \frac{1}{2}(i - \phi_u)f_{\overline{z}} + \frac{1}{2}\phi_{\overline{z}}f_u$$

and so (13) and (14) differ only by a factor of $-2i$.

For a hypersurface $M^{2n+1} \subset \mathbb{C}^{n+1}$ given as the graph

$$M = \{(z_1, \ldots, z_n, u + iv): v = \phi(z, \overline{z}, u)\}$$

we obtain n vector fields of the same form as in (14); namely,

$$L_j = (1 + i\phi_u)\frac{\partial}{\partial \overline{z}_j} - i\phi_{\overline{z}_j}\frac{\partial}{\partial u}, \qquad j = 1, \ldots, n.$$

These are clearly linearly independent and annihilate the restrictions of holomorphic functions. Thus they provide a basis for V.

We now consider the simplest and most important examples of the induced CR structure on hypersurfaces in $\mathbb{C}^2$. We use (z, w) as coordinates with $z = x + iy$ and $w = u + iv$. Two of our examples are graphs of the form $M = \{v = f(x, y, u)\}$:

1. $\mathbb{R}^3 = \{(x, y, u, v): v = 0\}$,

2. $Q = \{(x, y, u, v): v = x^2 + y^2\}$.

The third example is the sphere

3. $S = \{(x, y, u, v): x^2 + y^2 + u^2 + v^2 = 1\}$.

We also use the notations

$$Q = \{(z, w): v = |z|^2\},$$
$$S = \{(z, w): |z|^2 + |w|^2 = 1\}.$$

Q is called the *hyperquadric*. It is easy to picture since the slice of Q in each three-dimensional space $u = $ constant is just a paraboloid. Q appears in other areas of several complex variables as the Heisenberg group and the boundary of the Siegel upper half space.

EXAMPLE 1. $\mathbb{R}^3$. Since L has the form

$$L = \alpha\frac{\partial}{\partial z} + \beta\frac{\partial}{\partial \overline{w}}$$

and it must annihilate the defining function $r = \frac{1}{2i}(w - \overline{w})$, we can take

$$L = \frac{\partial}{\partial \overline{z}}.$$

We know that $H = \{\operatorname{Re} L, \operatorname{Im} L\}$, so $H = \{\frac{\partial}{\partial x}, \frac{\partial}{\partial y}\}$. Or, we can reason directly from the definition. For $M = \mathbb{R}^3$, $T_pM = \{\frac{\partial}{\partial x}, \frac{\partial}{\partial y}, \frac{\partial}{\partial u}\}$ and $JT_pM = \{\frac{\partial}{\partial x}, \frac{\partial}{\partial y}, \frac{\partial}{\partial v}\}$. So

$$H = T_pM \cap J(T_pM) = \left\{\frac{\partial}{\partial x}, \frac{\partial}{\partial y}\right\}.$$

Note that $[H, H] \subset H$.

EXAMPLE 2. Q. Here the defining function is $r = \frac{1}{2i}(w - \overline{w}) - |z|^2$ and so we may take for L,

$$L = \frac{\partial}{\partial \overline{z}} - 2iz\frac{\partial}{\partial \overline{w}}.$$

Or, if we use (14),

$$L = \frac{\partial}{\partial \overline{z}} - iz\frac{\partial}{\partial u}.$$

At the point $p = (x, y, u, x^2 + y^2) \in Q$, the characteristic plane is $H = \{\operatorname{Re} L, \operatorname{Im} L\} = \left\{\frac{\partial}{\partial x} + 2y\frac{\partial}{\partial u}, \frac{\partial}{\partial y}, -2x\frac{\partial}{\partial u}\right\}$. Note that $[H, H] \not\subset H$.

EXERCISE 7. Find this plane by using

$$H = T_pQ \cap J(T_pQ).$$

(Hint: Be careful to distinguish between, for instance, $\frac{\partial}{\partial x}$ as a vector in $\mathbb{R}^4$ and as a tangent vector to the graph of $v = x^2 + y^2$.)

Note that the curves $x = t$, $y = t$, $u = 0$ and $x = t$, $y = -t$, $u = 0$ are everywhere in H, in the sense that the tangent vector at each point p on the curve is in H_p. These curves, which have some interesting properties, will be studied in Chapter 9.

The vector field

$$L = \frac{\partial}{\partial \overline{z}} - iz\frac{\partial}{\partial u},$$

which generates V, is very well known in partial differential equations since it was the first example of a nonsolvable operator. The discovery by Hans Lewy [**Le2**] that the equation $Lu = f$ is not always solvable was a seminal event in the modern theory of partial differential equations. See [**Tr**] for an interesting discussion of its contemporaneous effect and lasting influence.

As we shall show in Chapter 10, the induced Cauchy–Riemann operator on any strictly pseudoconvex hypersurface in $\mathbb{C}^2$ is also nonsolvable. Further, the induced operators on hypersurfaces in $\mathbb{C}^{n+1}$ give rise to systems with various nonsolvability properties. In recognition of the profound influence of [**Le2**] the induced Cauchy–Riemann operators of submanifolds of $\mathbb{C}^{n+1}$ are often called *Lewy operators*. In Section 4 of the present chapter, we use this nonsolvability as the basis for another type of nonsolvability result and then continue the discussion in Chapter 10.

EXAMPLE 3. S^3. The defining function is $r = |z|^2 + |w|^2 - 1$. So

$$L = w\frac{\partial}{\partial \overline{z}} - z\frac{\partial}{\partial \overline{w}}$$

and the real and imaginary points give a basis for H:

$$v_1 = u \frac{\partial}{\partial x} - v \frac{\partial}{\partial y} - x \frac{\partial}{\partial u} + y \frac{\partial}{\partial v},$$

and

$$v_2 = v \frac{\partial}{\partial x} + u \frac{\partial}{\partial y} - y \frac{\partial}{\partial u} - x \frac{\partial}{\partial v}.$$

We can also compute H by making use of (8). The normal to S^3 at the point $Z = (z, w)$ is just $Z_{\mathbb{R}}$. So a vector Y is in H if it is orthogonal to $Z_{\mathbb{R}}$ and $JZ_{\mathbb{R}}$. The condition for this is

$$Y_{\mathbb{C}} \cdot \bar{Z} = 0.$$

So for $Y_{\mathbb{C}}$ we may take $(\bar{w}, -\bar{z})$ or $(i\bar{w}, -i\bar{z})$. These vectors are independent over $\mathbb{R}$ and so their real forms span H. We obtain that at the point $(x, y, u, v) \in S^3$,

$$H = \text{linear span } \{(u, -v, -x, y), (v, u, -y, -x)\}.$$

EXERCISE 8. Construct H in the following manner: Let $X \in S^3$ and use the Euclidean metric on $\mathbb{R}^4$ to obtain

$$H_X = \{Y : Y \perp X, \quad Y \perp JX\}.$$

Here is the higher-dimensional hyperquadric.

EXAMPLE 4. $Q_{p,q}$. In $\mathbb{C}^{n+1}$, let

$$Q_{p,q} = \left\{ (z_1, \ldots, z_{n+1}) : \text{Im } z_{n+1} = \sum_1^p |z_j|^2 - \sum_{p+1}^{p+q} |z_j|^2 \right\}.$$

As a basis for V we may take the operators

$$L_k = \frac{\partial}{\partial \bar{z}_k} - i z_k \frac{\partial}{\partial u} \qquad k = 1, \ldots, p,$$

$$L_j = \frac{\partial}{\partial \bar{z}_j} + i z_j \frac{\partial}{\partial u} \qquad j = p+1, \ldots, p+q.$$

It is a very useful fact that Q and S^3 are locally equivalent and, indeed, even globally equivalent after one point p is removed from S^3. See Lemma 2.7 for one biholomorphism on $\mathbb{C}^2 - \{p\}$ which effects this equivalence.

Another hypersurface which is globally equivalent to Q is the product of a parabola with $\mathbb{R}^2$ given by

$$v = y^2.$$

EXERCISE 9. Show that the transformation

$$z_1 = z, \qquad z_2 = w - iz^2$$

takes

$$\text{Im } w = |z|^2$$

to

$$\text{Im } z_2 = (\text{Im } z_1)^2,$$

and takes

$$L = \frac{\partial}{\partial \bar{z}} - iz \frac{\partial}{\partial u}$$

to

$$\hat{L} = \frac{\partial}{\partial \bar{z}_1} + y \frac{\partial}{\partial s}$$

where $\text{Im } z_1 = y$ and $\text{Im } z_2 = s$.

We have seen in Exercise 3 that $\mathbb{R}^3$ and Q are not locally equivalent under a biholomorphism of open sets of $\mathbb{C}^2$. It is easy to see this same fact in another way; for $\mathbb{R}^3$ we have $[H, H] \subset H$ while for Q, $[H, H] \not\subset H$. Note that the condition $[H, H] \subset H$ on some M^{2n+1} implies that M is foliated by codimension-one submanifolds which are everywhere tangent to the $2n$-plane distribution H. (This is one form of the Frobenius theorem. See Chapter 5.) Further, the J operator on H provides a complex structure on each N. This, of course, is clear for $\mathbb{R}^3$. It is important to realize that the condition $[V, V] \subset V$, which we require for all CR manifolds, says nothing about the existence of special submanifolds. This is because V consists of complex vector fields and the Frobenius theorem does not apply. (Note, however, that we shall soon be making use of a holomorphic version of the Frobenius theorem. This is a different matter and is no more difficult than the usual version.)

It is natural to focus attention on the general case where there is no foliation of M by complex hypersurfaces. So we take the distribution H to be nonintegrable, i.e.,

$$[H, H] \not\subset H.$$

Indeed we will take H as far from integrable as possible. We express this in terms of V. So let $\{L_1, \dots, L_n\}$ be a basis for V, near some point x, and choose some real vector U transverse to H. Thus $\{L_1, \dots, L_n, \bar{L}_1, \dots, \bar{L}_n, U\}$ is a basis for $\mathbb{C} \otimes TM$ near x. Let

$$[L_j, \bar{L}_k] = i g_{jk} U \qquad (\text{mod } \{L_1, L_2, \dots, \bar{L}_n\})$$

define the matrix g. This matrix is often referred to as the Levi form. It is clearly Hermitian. Although g itself depends on a choice of basis, its most important property does not. Let us say that an Hermitian matrix has signature (p, q, r) if it has p positive eigenvalues, q negative eigenvalues, and r zero eigenvalues, except that we use (p, q) when $r = 0$. Further let us identify (p, q, r) with (q, p, r). Then the signature of the Levi form does not depend on the choices. To see this, we make a second choice $\{\tilde{L}_1, \dots, \tilde{L}_n, \tilde{U}\}$ with

$$\tilde{L}_j = A_{jk} L_k$$

and

$$\tilde{U} = b\,U \qquad (\text{mod } \{L, \overline{L}\})$$

with A invertible and b real and nonzero. Then

$$\tilde{g} = b^{-1} A\, g\, A^*$$

and so $\tilde{g}$ has the same signature as g.

DEFINITION. M^{2n+1} is *nondegenerate* at x if $g(x)$ is a nonsingular matrix.

In this case $p + q = n$.

DEFINITION. M^{2n+1} is *strictly pseudoconvex* at x if $g(x)$ is either positive definite or negative definite.

Here g has signature $(n, 0)$ $\big(\text{or } (0, n)\big)$.

EXAMPLES. Q and S are nondegenerate and $Q_{(p,q)}$ has signature $(p, q, n - p - q)$.

REMARKS.

1. On M^3 the concepts of nondegenerate and strictly pseudoconvex coincide. Further, in this dimension, they are equivalent to H being nonintegrable.

2. To rephrase a previous observation: $\mathbb{R}^3$ and Q (or S^3) are not locally equivalent since Q (or S^3) is everywhere nondegenerate and $\mathbb{R}^3$ is nowhere nondegenerate.

Our main concern will be with nondegenerate submanifolds $M^3 \subset \mathbb{C}^2$. We shall see that a great deal can be accomplished by modelling such hypersurfaces on the hyperquadric. Here is a first step.

LEMMA 4. *If M^3 is nondegenerate at p, then there is a complex affine map $\mathscr{A}$ which takes p to the origin and such that $\mathscr{A}(M)$ has the form*

$$(15) \qquad v = a|z|^2 + Az^2 + \overline{A}\overline{z}^2 + bu^2 + Bzu + \overline{B}\overline{z}u + \cdots$$

with $a \neq 0$.

PROOF. We first translate p to the origin. We then choose a point R on H_0 and a point S not on H_0 and find the complex linear map which takes R to $(1, 0)$ and S to $(0, 1)$. We end up with a map $\mathscr{A}_1$ so that $\mathscr{A}_1(H_0)$ is the z-axis. We now take a linear map of the form $(z, w) \to (z, e^{i\phi}w)$ which takes $T_0\mathscr{A}_1(M)$ to the (z, u)-plane. The composition map $\mathscr{A}$ takes M to the form (15) and we need only show that a is nonzero. We see from (14) that $[L, \overline{L}]$ has the form

$$[L, \overline{L}] = L(1 - i\phi_u)\frac{\partial}{\partial z} + L(i\phi_z)\frac{\partial}{\partial u}$$

$$- \overline{L}(1 + i\phi_u)\frac{\partial}{\partial \overline{z}} - \overline{L}(-i\phi_{\overline{z}})\frac{\partial}{\partial u}$$

and so, using (15), at the origin

$$[L, \overline{L}] = 2ia\frac{\partial}{\partial u} - i\overline{B}\,\overline{L} - iBL.$$

Thus the fact that M, and so also $\mathscr{A}(M)$, is nondegenerate implies that $a \neq 0$.

§3. **CR functions.** We usually will be concerned with local questions. When this needs to be emphasized we write (M, V, p) to denote the CR structure on M in the neighborhood of one of its points p. The notation (M, p) refers to a neighborhood of p in the manifold M and is used also when M does not have a CR structure. For example compare these two definitions.

DEFINITION. A complex-valued function f on M is a *CR function* on (M, V) if $Lf = 0$ for all sections $L: M \to V$.

DEFINITION. A complex-valued function f on a neighborhood of $p \in M$ is a *CR function* on (M, V, p) if there exists another neighborhood U_p such that $Lf = 0$ for all sections $L: U_p \to V$.

Just as the best example of a CR manifold is a hypersurface in $\mathbb{C}^n$, the best example of a CR function is a holomorphic function restricted to such a hypersurface.

LEMMA 5. *Let M^{2n+1} be a hypersurface in $\mathbb{C}^{n+1}$, p a point of M, U a neighborhood of p in $\mathbb{C}^{n+1}$ and $U_p = M \cap U$ the corresponding neighborhood of p in M. If F is holomorphic on U then the restriction of F to U_p is a CR function on (M, p).*

PROOF. This is essentially the definition of the induced structure. For let f denote this restriction. Each $L \in V$ is the restriction to M of a vector field in $H_{0,1}(\mathbb{C}^{n+1})$. So, $LF = 0$ and, since L is tangent to M, $Lf = 0$.

The converse is certainly false. Not every CR function on a hypersurface is the restriction of some holomorphic function.

EXERCISE 10.

a. Find a function $F(z)$ holomorphic on $\{z \in \mathbb{C}: |z| < 1\}$, smooth up to $S = \{z \in \mathbb{C}: |z| = 1\}$ but not the restriction of any function holomorphic on a larger set.

b. Construct from F a CR function on S^3 which is not the restriction of a function holomorphic in a neighborhood of any point of (z, w), where $|z| = 1$ and $w = 0$.

A simple limiting argument provides this important variation of the lemma.

COROLLARY. *Let M, p, and U be as above and let F be holomorphic on one component of $U - (M \cap U)$. If F extends smoothly to M, then this extension is a CR function on U_p.*

EXAMPLE. There exist functions holomorphic in $\{(z, w): |z|^2 + |w|^2 < 1\}$ and smooth up to the boundary with the function restricted to the boundary in C^∞ but not real analytic. Thus the restriction is a CR function on S^3 which does not extend to a holomorphic function.

EXAMPLE. If we permit M to be degenerate, then it is very easy to find explicit CR functions on M which are not the boundary values of holomorphic functions. For instance, any function $f(u)$ is a CR function on $\mathbb{R}^3$.

However, when the hypersurface and the CR function are both real analytic, then holomorphic extensions do exist. Not surprisingly, in this case the extension can be given almost explicitly.

THEOREM 1. *If M is a real analytic hypersurface in $\mathbb{C}^{n+1}$ and f is a real analytic, complex-valued, CR function on (M, p) then f is the restriction to M of a function holomorphic in a neighborhood of p.*

PROOF. We may take p to be the origin and M to be of the form

$$v = \phi(z, \overline{z}, u)$$

where $z = (z_1, \ldots, z_n)$, $z_{n+1} = u + iv$, and ϕ is real analytic with $d\phi(0) = 0$. We have seen that a basis for V is

$$(16) \quad L_j = \left(1 + i\phi_u(z, \overline{z}, u)\right) \frac{\partial}{\partial \overline{z}_j} - i\phi_{\overline{z}_j}(z, \overline{z}, u) \frac{\partial}{\partial u}, \qquad j = 1, \ldots, n.$$

We are seeking a holomorphic function F with

$$F\left(z, u + i\phi(z, \overline{z}, u)\right) = f(z, \overline{z}, u).$$

We complexify u to obtain

$$F\left(z, \zeta + i\phi(z, \overline{z}, \zeta)\right) = f(z, \overline{z}, \zeta).$$

In other words, the value at the point (z, z_{n+1}) of the extension, if this extension exists, is $f(z, \overline{z}, \zeta)$ where ζ satisfies

$$(17) \qquad \qquad \zeta + i\phi(z, \overline{z}, \zeta) = z_{n+1}.$$

So to prove the theorem, let us define ζ by (17) and define F by

$$F = f\left(z, \overline{z}, \zeta(z, \overline{z}, z_{n+1})\right).$$

This function is holomorphic in z_{n+1}. The question is, is it holomorphic in each z_j for $1 \le j \le n$.

From (17) we see that

$$\frac{\partial \zeta}{\partial \overline{z}_j} = \frac{-i\phi_{\overline{z}_j}(z, \overline{z}, \zeta)}{1 + i\phi_\zeta(z, \overline{z}, \zeta)},$$

and so

$$\frac{\partial F}{\partial \overline{z}_j} = \frac{\partial f}{\partial \overline{z}_j} + \frac{\partial f}{\partial \zeta} \frac{\partial \zeta}{\partial \overline{z}_j}$$

$$= \frac{\partial f}{\partial \overline{z}_j} - \frac{i\phi_{\overline{z}_j}}{1 + i\phi_\zeta} \frac{\partial f}{\partial \zeta}$$

where the right-hand side is evaluated at $(z, \overline{z}, \zeta)$ with ζ given by (17). But $L_j f = 0$ implies

$$\left(1 + i\phi_\zeta(z, \overline{z}, \zeta)\right) \frac{\partial f}{\partial \overline{z}_j} - i\phi_{\overline{z}_j}(z, \overline{z}, \zeta) \frac{\partial f}{\partial \zeta} = 0$$

since L_j and f are real analytic. So $\frac{\partial F}{\partial \overline{z}_j} = 0$, and we are done.

We conclude with alternative characterizations of CR functions which foreshadow the definition of a CR map. Given a map $\phi: M \to N$ we denote the induced map $TM \to TN$ (or $\mathbb{C} \otimes TM \to \mathbb{C} \otimes TN$) by $d\phi$ or ϕ_*.

LEMMA 6. *The following conditions are equivalent.*

(a) *f is a CR function on (M, p).*

(b) *Near p, $f_* L = \lambda \frac{\partial}{\partial \overline{z}}$ for all L in V, where the function λ depends on L.*

(c) *For all $L \in V$, $df \cdot J_1 L = J_2 \cdot df L$ where J_1 gives the CR structure on M and J_2 the complex structure on $\mathbb{C}$.*

(d) *The same as* (c) *but with V replaced by H.*

PROOF. For any function and any $L \in \mathbb{C} \otimes TM$,

$$f_*(L) = L(f)\frac{\partial}{\partial z} + L(\overline{f})\frac{\partial}{\partial \overline{z}}.$$

Condition (a) means that, for all $L \in V$,

$$L(f) = 0.$$

So (a) and (b) are clearly equivalent. Condition (c) is equivalent to the fact that, for all $L \in V$,

$$(-i)df L = J_2(df L),$$

and so $df L \in \{\frac{\partial}{\partial \overline{z}}\}$. Thus (b) and (c) are also equivalent. We have $X \in H$ if and only if $X - iJX \in V$. The equivalence of (c) and (d) follows from considering real and imaginary parts.

§4. **CR maps and the realization problem.** We will use the notation $F: (M_1, p) \to (M_2, q)$ to mean that F is a map of an open set in one manifold M_1 into another manifold M_2 with $F(p) = q$. Just as we formulated a global and a local definition of a CR function, we can give two definitions of CR mappings.

DEFINITION. A map $F: M_1 \to M_2$ is a *CR map*, with respect to the CR structures (M_1, V_1) and (M_2, V_2), if $F_* V_1 \subset V_2$. We write $F: (M_1, V_1) \to (M_1, V_2)$.

DEFINITION. A map $F: (M_1, p) \to (M_2, q)$ is a *CR map* with respect to the CR structures (M_1, V_1, p) and (M_2, V_2, q) if there exists a neighborhood U_p such that $F_* V_1 \subset V_2$ at all points of U_p. We write $F: (M_1, V_1, p) \to (M_2, V_2, q)$.

If F is also a diffeomorphism then naturally we say that it is a *CR diffeomorphism* and that M_1 and M_2 (or (M_1, p) and (M_2, q)) are CR diffeomorphic. For brevity we sometimes just say they are equivalent.

EXERCISE 11. Show that F is a CR map if and only if $dF \cdot J_1 = J_2 \cdot dF$.

DEFINITION. A CR structure (M, V) is *realizable* if there exists some hypersurface M_1 in $\mathbb{C}^{n+1}$ such that M is equivalent to M_1, where M_1 has the induced CR structure.

DEFINITION. A CR structure (M, V, p) is *realizable* if there exists some hypersurface M_1 in $\mathbb{C}^{n+1}$ such that (M, p) is equivalent to (M_1, q) where M_1 has the induced CR structure.

LEMMA 7. *Let* $F: M_1 \to M_2 \subset \mathbb{C}^{n+1}$ *be a realization of* M_1. *Write* $F = (f_1, \ldots, f_{n+1})$. *Then each* f_j *is a CR function.*

PROOF. For any $F: M_1 \to \mathbb{C}^{n+1}$ and any $L \in \mathbb{C} \otimes TM_1$ we have

$$F_*(L) = \sum_{j=1}^{n+1} \left\{ L(f_j) \frac{\partial}{\partial z_j} + L(\bar{f}_j) \frac{\partial}{\partial \bar{z}_j} \right\},$$

and so

$$F_*: V_1 \to V_2 \subset \left\{ \frac{\partial}{\partial \bar{z}_1}, \ldots, \frac{\partial}{\partial \bar{z}_n} \right\}$$

implies that

$$L(f_j) = 0.$$

We have an obvious converse of this.

LEMMA 8. *If* $f_j: M \to \mathbb{C}$ *is CR for* $j = 1, \ldots, n+1$ *and if for* $F = (f_1, \ldots, f_n)$ *we have that* $F(M)$ *is a smooth hypersurface and that* $F: M \to F(M)$ *is a diffeomorphism, then* M *is equivalent to* $F(M)$ *and so is realizable.*

There is also a somewhat less obvious converse. To discuss it we first recall a special case of a result from advanced calculus.

LEMMA 9. *Let* $\Phi: (M^m, p) \to \mathbb{R}^{m+1}$ *have components* $\Phi = (u_1, \ldots, u_{m+1})$. *These two conditions are equivalent.*

   (a) *There exists a neighborhood* U_p *of* p *for which* $M' = \Phi(U_p)$ *is a smooth submanifold of* $\mathbb{R}^{m+1}$ *and* $\Phi: U_p \to M'$ *is a diffeomorphism.*

   (b) *There exists an index* j *such that at* p

$$du_1 \wedge \cdots \wedge \widehat{du_j} \wedge \cdots \wedge du_{m+1} \neq 0$$

(*the notation means the m-form with* du_j *omitted*).

This can be reformulated for mappings into $\mathbb{C}^{n+1}$. For in this case (b) becomes, after possible relabeling,

(18) $$df_1 \wedge d\bar{f}_1 \wedge \cdots \wedge df_n \wedge d\bar{f}_n \wedge df_{n+1} \neq 0.$$

This has nothing to do with CR structures. The image $F(M^{2n+1})$ is smooth if and only if (18) holds (after a renumbering). The result we are seeking replaces (18) by a weaker condition in the case of CR maps.

LEMMA 10. (M, p) *is realizable if and only if there exist functions* $f_1, \ldots,$ f_{n+1} *such that on some neighborhood of* p

(19) $$Lf_k = 0 \quad \text{for } LV \text{ and all } k$$

and

(20) $$df_1 \wedge \cdots \wedge df_{n+1} \neq 0.$$

PROOF. If (M, p) is realizable then it may be identified with some hypersurface

$$\{(z_1, \ldots, z_n, u + iv): v = \phi(z, \overline{z}, u)\}.$$

Let $f_k = z_k$ for $k = 1, \ldots, n$, and let $f_{n+1} = u + i\phi(z, \overline{z}, u)$. Each f_j is the restriction to M of a holomorphic function and so (19) holds. Further

$$df_1 \wedge \cdots \wedge df_{n+1} = (1 + i\phi_u) \, dz_1 \wedge \cdots \wedge dz_n \wedge du \quad (\text{mod } d\overline{z})$$

and so (20) holds.

Conversely, if functions $f_1, \ldots, f_{n+1}$ satisfy (19) then $F = (f_1, \ldots, f_{n+1})$ satisfies $F_* V \subset H_{0,1}(\mathbb{C}^{n+1})$ so we only have to prove that (20) implies that $F(M)$ is smooth. That is, we need to show that (20) together with (19) implies (18). So we start with a basis $\{L_1, \ldots, L_n\}$ for V and complete to a basis $\{L_1, \ldots, L_n, \overline{L}_1, \ldots, \overline{L}_n, U\}$ for $\mathbb{C} \otimes TM$. From (20) we see that

$$df_1 \wedge \cdots \wedge df_{n+1}(\overline{L}_1, \ldots, \overline{L}_n, U) \neq 0$$

and so we may renumber the functions and assume that

(21) $$df_1 \wedge \cdots \wedge df_n(\overline{L}_1, \ldots, \overline{L}_n) \neq 0.$$

We claim that

$$df_1 \wedge \cdots \wedge df_n \wedge d\overline{f}_1 \wedge \cdots \wedge d\overline{f}_n \wedge df_{n+1} \neq 0.$$

For if the left-hand side is zero at a point then

(22) $$df_{n+1} = \sum_{}^{n} \alpha_j \, df_j + \sum_{}^{n} \beta_k \, d\overline{f}_k$$

and β_k would satisfy the linear system

$$\sum_{}^{n} \beta_k \, d\overline{f}_k(L_j) = 0.$$

However, (21) states that the determinant of this matrix is nonzero. Thus we would have that each β_k is zero. But then (22) would contradict (20).

This lemma tells us that the realizability problem for CR manifolds is also a natural question about partial differential equations. Given k first-order partial differential operators $L_1, \ldots, L_k$ on $\mathbb{R}^N$ which satisfy the compatibility condition

$$[L_i, L_j] \in \{L_1, \ldots, L_k\},$$

we may ask if the over-determined system

$$L_j f = 0, \quad j = 1, \ldots, k,$$

has $N - k$ independent solutions. This is so for real analytic operators. When $k = n$, $N = 2n + 1$, and $\{L_1, \ldots, \overline{L}_n\}$ is linearly independent, this system gives the realization problem for the CR structure defined by $V =$ linear span $\{L_1, \ldots, L_n\}$. In particular, real analytic CR structures are realizable. We now give two other direct proofs of this fact.

THEOREM 2. *Any real analytic CR manifold is locally realizable.*

In each of our proofs we start with a basis for V over some set $U \subset M$,

$$L_j = \sum_1^{2n+1} \alpha_{jk}(x) \frac{\partial}{\partial x_k}, \qquad j = 1, \ldots, n,$$

with α_{jk} a real analytic function of $x \in U$. Of course, we have the compatibility condition

$$[L_j, L_k] = \sum_{r=1}^n \beta_{jkr}(x) L_r.$$

PROOF 1. The complexifications

$$M_j = \sum^{2n+1} \alpha_{jk}(z) \frac{\partial}{\partial z_k}, \qquad j = 1, \ldots, n,$$

are holomorphic vector fields and satisfy

$$[M_j, M_k] = \sum^n \beta_{jkr}(z) M_r.$$

By the Frobenius theorem for holomorphic vector fields (see Chapter 5) there exist holomorphic functions $f_1(z), \ldots, f_{n+1}(z)$ with

$$M_j f_k = 0 \quad \text{and} \quad df_1 \wedge \cdots \wedge df_{n+1} \neq 0.$$

So the functions $g_k(x)$ obtained by setting Im $z = 0$ satisfy $L_j g_k = 0$ and $dg_1 \wedge \cdots \wedge dg_{n+1} \neq 0$ and so give a local CR embedding of M into $\mathbb{C}^{n+1}$.

PROOF 2. Here we complexify only one variable. We may assume that

$$\frac{\partial}{\partial x_{2n+1}} \notin V \oplus \overline{V}.$$

We take

$$M_j = \sum \alpha_{jk}(x_1, \ldots, x_{2n}, x_{2n+1} + it) \frac{\partial}{\partial x_k}$$

and define a new operator

$$M_{n+1} = \frac{\partial}{\partial x_{2n+1}} + i \frac{\partial}{\partial t}.$$

For $V_0 =$ linear span $\{M_1, \ldots, M_{n+1}\}$ we still have $[V_0, V_0] \subset V_0$ so the almost complex structure defined by V_0 is integrable and M is given as the hypersurface $\{t = 0\}$ in this complex structure.

Not all CR structures are realizable. Subtle questions of partial differential equations are involved. The problem was first formulated by Hans Lewy

[Le1] and examples of nonrealizable structures were first given by Nirenberg ([Ni2]). These examples are three-dimensional, strictly pseudoconvex CR structures. In Chapter 10 we discuss a generalization and simplification of Nirenberg's examples. For higher dimensions, it is known that not all CR structures of signature $(n - 1, 1)$ are realizable ([JT2]) and that all strictly pseudoconvex structures of dimension at least seven are locally realizable ([Ku], [Ak], [We3]). The latter result is the most difficult in the subject. For dimension five the realizability of strictly pseudoconvex structures is still open. Recently D. Catlin [Cat] has announced a new proof of realizability of strictly pseudoconvex structures of dimension at least seven and the extension of the result to certain other signatures. There are also counterexamples for degenerate structures, i.e., for structures where the Levi form has a zero eigenvalue. For instance, there is the very interesting result of LeBrun ([LeB]) relating realizability to the analyticity of an associated conformal metric and the generalization of this result by Rossi ([Ros2]).

Here is one very simple nonrealizable example for a degenerate structure. In a somewhat different form, it is due to Hill, Penrose, and Sparling [Pe], see also [Ea]. We have pointed out that

$$L = \frac{\partial}{\partial \overline{z}} - iz\frac{\partial}{\partial u}$$

is nonsolvable. So let $g(z, \overline{z}, u)$ be a C^∞ function for which $Lu = g$ has no solution in any neighborhood of the origin in $\mathbb{R}^3$. Let M be an open neighborhood of the origin in $\mathbb{R}^5$ on which we take coordinates (z, u, ζ). Let

$$L_1 = L - g(z, \overline{z}, u)\zeta\frac{\partial}{\partial \zeta},$$

$$L_2 = \frac{\partial}{\partial \overline{\zeta}}.$$

These operators define a CR structure of signature $(1, 0)$. We claim this structure is not realizable. Note that each function f annihilated by L_1 and L_2, i.e., each CR function f, must satisy $f_{\overline{z}}(0) = f_{\overline{\zeta}}(0) = 0$. Thus for M to admit three independent CR functions there would have to be some CR function f with $f_\zeta(0) \neq 0$.

LEMMA 11. *Every function satisfying $L_1 f = 0$ on a neighborhood of the origin has $f_\zeta(0) = 0$.*

PROOF. Assume to the contrary that $L_1 f = 0$ and $f_\zeta(0) \neq 0$. Pick some branch of the logarithm and set $F(z, \overline{z}, u) = \ln f_\zeta(z, \overline{z}, u, 0, 0)$. So F is smooth near the origin. From

$$Lf - g\zeta f_\zeta = 0$$

we derive

$$Lf_\zeta - gf_\zeta - g\zeta f_{\zeta\zeta} = 0,$$

and so, at $\zeta = 0$,

$$Lf_\zeta - gf_\zeta = 0.$$

Hence

$$LF - g = 0.$$

But our choice of g makes this impossible. A somewhat different proof along with a generalization to higher dimensions may be found in [**Ja3**].

§5. **CR structures by means of differential forms.** In E. Cartan's approach to differential geometry, the basic object is differential forms rather than vector fields. So we need to reformulate our discussion of CR structures. For the elementary properties of differential forms and calculations with exterior differentiation see, for example, [**Ca4**], [**KN**], or [**Sp**].

Given (M, V, p) we first choose some real nonzero form ω which annihilates H. We may write this in various ways: $\omega^\perp = H$; $\omega \in H^\perp$; $\omega X = 0$ for $X \in H$; or $\omega(L) = 0$ for $L \in V \otimes \overline{V}$.

Next we choose $\omega_1, \ldots, \omega_n$ so that

$$\text{lin span } \{\omega, \omega_1, \ldots, \omega_n\} = V^\perp.$$

In a neighborhood of p, we now have a real form ω and complex forms $\omega_1, \ldots, \omega_n$ with

$$\omega \wedge \omega_1 \wedge \cdots \wedge \omega_n \wedge \overline{\omega}_1 \wedge \cdots \wedge \overline{\omega}_n \neq 0.$$

Sometimes we write ω_o for ω and use the notation

$$\{\omega_j : 0 \leq J \leq n\},$$

rather than

$$\{\omega_j : 1 \leq j \leq n\}.$$

Any other choices on (M, V, p) would have to satisfy

$$\widetilde{\omega} = a\omega, \qquad \widetilde{\omega}_j = \sum_{k=1}^{n} A_{jk}\omega_k + B_j\omega$$

with a real and nonzero and (A_{jk}) invertible.

The space lin span $\{\omega, \omega_1, \ldots, \omega_n\}$ is thus well-defined. It is often denoted by $H^{0,1}(M)$. Thus $H_{0,1}(M)$ is a subspace of $\mathbb{C} \otimes T(M)$ and $H^{0,1}(M)$ is a subspace of $\mathbb{C} \otimes T^*(M)$. In this book we will not use the other spaces $H_{p,q}$ and $H^{p,q}$, except for $H_{1,0} = \overline{H_{0,1}}$.

The integrability condition $[V, V] \subset V$ has a natural formulation in terms of these differential forms. Let B be a sub-bundle of TM or of $\mathbb{C} \otimes TM$. In the first case define

$$B^\perp = \{\omega \in T^*M : \omega(X) = 0 \text{ for } X \in B\},$$

and in the second case

$$B^\perp = \{\omega \in \mathbb{C} \otimes T^*M : \omega(X) = 0 \text{ for } X \in B\}.$$

LEMMA 12. *The following are equivalent*:

$$[B, B] \subset B$$

and

$$d\omega \equiv 0 \pmod{B^{\perp}} \quad \text{for all } \omega \in B^{\perp}.$$

EXERCISE 12. Provide the proof. (Hint: A good place to start is the equation

$$d\theta(X, Y) = X\theta(Y) - Y\theta(X) - \theta[X, Y],$$

which is valid for all 1-forms θ and vector fields X and Y.)

Thus starting from (M, V, p), we obtain forms $\{\omega, \omega_1, \ldots, \omega_n\}$ with the properties

$$\omega \text{ is real},$$

$$\omega \wedge \omega_1 \wedge \cdots \wedge \omega_n \wedge \overline{\omega}_1 \wedge \cdots \wedge \overline{\omega}_n \neq 0,$$

and

$$(23) \qquad d\omega_J \equiv 0 \bmod \{\omega_0, \ldots, \omega_n\}, \qquad J = 0, \ldots, n.$$

Conversely, given such forms in a neighborhood of p, we define (M, V, p) by setting $V = \{\omega, \omega_1, \ldots, \omega_n\}^{\perp}$.

REMARK. On M^3 we have $\{\omega, \omega_1, \overline{\omega}_1\}$ so for any 1-form θ we have

$$d\theta = \alpha\omega\omega_1 + \beta\omega\overline{\omega}_1 + a\omega_1\overline{\omega}_1 \equiv 0 \bmod \{\omega, \omega_1\}$$

and, as we have seen in §2, there is no integrability restriction for dimension 3.

It is easy to see what these forms are when M^{2n+1} is realized in $\mathbb{C}^{n+1}$. First let us take M given by an arbitrary defining function

$$(24) \qquad M = \{z \in \mathbb{C}^{n+1} : r(z, \overline{z}) = 0\}.$$

We claim that for ω we may take

$$(25) \qquad \omega = i\partial r.$$

We must show that this form is real when evaluated on vectors tangent to M and that it annihilates vectors in H. We have

$$dr = \partial r + \overline{\partial} r,$$

so

$$\partial r + \overline{\partial} r = 0$$

when evaluated on $X \in TM$. Thus $i\partial r(X)$ is real. Further, if $X \in H$ then $X + iJX \in V \subset H_{0,1}(\mathbb{C}^{n+1})$, and so

$$\partial r(X + iJX) = 0$$

from which follows

$$\partial r X = 0.$$

EXAMPLE. For S^3, we have

$$r = |z_1|^2 + |z_2|^2 - 1,$$

so

$$\omega = i\partial r = i(\bar{z}_1 dz_1 + \bar{z}_2 dz_2).$$

Let $p = (z, w) = (x + iy, u + iv) \in S^3$ and let $U \in T_p S^3$. So

$$U(\bar{z}_j)z_j + U(z_j)\bar{z}_j = 0,$$

and $U(z_j)\bar{z}_j$ is imaginary.

$$\omega(U) = iU(z_j)\bar{z}_j$$

is real. Further, we know from Section 2 that H is spanned by $\{(\bar{z}_2, -\bar{z}_1),$
$(i\bar{z}_2, -i\bar{z}_1)\}$ so for $U \in H$, $\omega(U) = 0$.

We now use the notation

$$z = (z_1, \ldots, z_n) \quad \text{and} \quad z_{n+1} = w = u + iv,$$

and in place of (24) we take

(26) $M = \{z \in \mathbb{C}^{n+1} : v = f(z, \bar{z}, u)\}.$

So now

(27) $r = -\frac{1}{2i}(w - \overline{w}) + f(z, \bar{z}, u),$

and

(28) $\omega = i(f_{z_k} dz_k + \frac{1}{2}f_u dw) - \frac{1}{2}dw.$

But

$$dw = du + idf = (1 + if_u)du + if_{z_k} dz_k + if_{\bar{z}_k} d\bar{z}_k,$$

and so

(29) $\omega = -\frac{1}{2}(1 + f_u^2)du + \frac{1}{2}(if_{z_k} - f_u f_{z_k})dz_k + \frac{1}{2}(-if_{\bar{z}_k} - f_u f_{\bar{z}_k})d\bar{z}_k.$

The restrictions to M of the forms $dz_1, \ldots, dz_n$ annihilate V (since they annihilate all of $H_{0,1}(\mathbb{C}^{n+1})$) and satisfy

$$\omega \wedge dz_1 \wedge \cdots \wedge dz_n \neq 0.$$

So we may take $\omega_j = dz_j$, $j = 1, \ldots, n$.

For the integrability condition we need only show

(30) $d\omega \equiv 0 \bmod \{\omega, dz_1, \ldots, dz_n\}.$

We see from (28) that

$$dw \equiv 0 \bmod \{\omega, dz_1, \ldots, dz_n\}$$

and so when we compute $d\omega$ from (28) we do indeed get (30).

EXERCISE 13. Show directly that (29) implies (30).

On the hyperquadric in $\mathbb{C}^2$ we have that $f(z, \bar{z}, u) = |z|^2$ and so the CR structure is given by

(31)
$$\begin{cases} \omega = -\frac{1}{2}du + \frac{1}{2}i\bar{z}dz - \frac{1}{2}izd\bar{z}, \\ \omega_1 = dz. \end{cases}$$

EXAMPLE. We know that on the hyperquadric V is spanned by $L = \frac{\partial}{\partial\bar{z}} - iz\frac{\partial}{\partial u}$. We can see this again by computing $\{\omega, \omega_1\}^\perp$. We start with $L = \alpha\frac{\partial}{\partial z} + \beta\frac{\partial}{\partial\bar{z}} + \gamma\frac{\partial}{\partial u}$. From $\omega_1(L) = 0$, we see that $\alpha = 0$; then from $\omega(L) = 0$, we see that $\gamma + iz\beta = 0$. So $\{\omega, \omega_1\}^\perp$ is indeed spanned by $L = \frac{\partial}{\partial\bar{z}} - iz\frac{\partial}{\partial u}$. We have also considered the curve $z = t(1 + i)$, $u = 0$. We see again that this curve is in H:

$$\omega\left((1 + i)\frac{\partial}{\partial z} + (1 - i)\frac{\partial}{\partial\bar{z}}\right)$$
$$= \frac{1}{2}it(1 - i)(1 + i) - \frac{1}{2}it(1 + i)(1 - i)$$
$$= 0.$$

If we replace ω by $-\omega$ in (31) we have

(32)
$$d\omega = i\omega_1\bar{\omega}_1.$$

We can also obtain this, modulo $\{\omega\}$, for any strongly pseudoconvex $M^3 \subset \mathbb{C}^2$. We start with a more general result. Let $\{\omega, \omega_1, \dots, \omega_n\}$ define some CR structure. From the integrability condition in (23) we see that

$$d\omega \equiv h_{jk}\omega_j\bar{\omega}_k \mod \{\omega\}.$$

Since ω is real, $h^* = -h$. So we rewrite this as

(33)
$$d\omega \equiv ig_{jk}\omega_j\bar{\omega}_k \mod \{\omega\}$$

with g Hermitian.

LEMMA 13. *The CR structure is strongly pseudoconvex if and only if g is either positive or negative definite.*

PROOF. Choose some real vector U with $\omega(U) = 1$. Note that U is transverse to H. For each $L \in V$, define $a = a(L)$ by

$$[L, \bar{L}] \equiv iaU \mod V \oplus \bar{V}.$$

We know that the structure is strongly pseudoconvex precisely when $a(L)$ is nonzero for each $L \in V$. Now, since $\omega(L) = \omega(\bar{L}) = 0$, we see that $\omega[L, \bar{L}] = ia$. And, since $\omega_j(L) = \bar{\omega}_j(\bar{L}) = 0$, we may write this as

$$ia = -d\omega(L, \bar{L}) + L\omega(\bar{L}) + \bar{L}\omega(L)$$
$$= ig_{jk}(\omega_j(\bar{L}))(\bar{\omega}_k(L)).$$

So the structure is strongly pseudoconvex precisely when g is a definite matrix. By the appropriate choice of $\pm\omega$, we can obtain that g is positive definite. We then can replace each ω_j by some linear combination to obtain that g is the identity. For M^3, we can also obtain this by multiplying ω by some factor. This is convenient since it allows us to keep that ω_1 is dz.

LEMMA 14. *Let* $\{\omega, \omega_1\}$ *define a strongly pseudoconvex CR structure near* $p \in M^3$. *There is a nonzero function* ρ *for which the CR structure* $\{\theta, \omega_1\}$, *with* $\theta = \rho\omega$, *satisfies*

$$d\theta \equiv i\omega_1\overline{\omega}_1 \bmod\{\theta\}.$$

PROOF. For $\rho = g^{-1}$ we have

$$d\theta = \rho\, d\omega + \omega\, d\rho$$
$$\equiv \rho i g \omega_1 \overline{\omega}_1 \bmod \{\theta\}$$
$$\equiv i\omega_1\overline{\omega}_1 \bmod \{\theta\}.$$

We can eliminate the modulo term if we are willing to change ω_1 away from dz.

LEMMA 15. *A strongly pseudoconvex CR structure* (M^3, V, p) *can be defined by forms* $\{\omega, \omega_1\}$ *that satisfy*

$$d\omega = i\omega_1\overline{\omega}_1.$$

PROOF. Take any forms $\{\theta, \theta_1\}$ that give the structure. So

$$d\theta \equiv i g \theta_1 \overline{\theta}_1 \bmod \{\theta\},$$

with g nonzero. Let $\omega = g^{-1}\theta$. Then

$$d\omega = i\theta_1\overline{\theta}_1 + \alpha\omega\theta_1 + \overline{\alpha}\omega\,\overline{\theta}_1$$
$$= i(\theta_1\overline{\theta}_1 + \beta\omega\theta_1 - \overline{\beta}\omega\overline{\theta}_1)$$
$$= i(\theta_1 - \overline{\beta}\omega)(\overline{\theta}_1 - \beta\omega).$$

So we take $\omega_1 = \theta_1 - \overline{\beta}\omega$.

LEMMA 16. *Let* $\{\omega, \omega_1\}$ *give a CR structure on* M *and satisfy* $d\omega = i\omega_1\overline{\omega}_1 \bmod\{\omega\}$. *If* $\{\theta, \theta_1\}$ *defines the same CR structure and also satisfies* $d\theta \equiv i\theta_1\overline{\theta}_1 \bmod\{\theta\}$ *then there are functions* λ *and* μ, *with* $\lambda \neq 0$, *such that*

$$\theta = |\lambda^2|\omega,$$
$$\theta_1 = \lambda(\omega_1 + \mu\omega).$$

It is also easy to use differential forms to characterize CR functions and maps. Recall that $f: (M, V, p) \to \mathbb{C}$ is a CR function if $L(f) = 0$ for all $L \in V$ and $F: (M, V, p) \to (M_1, V_1, q)$ is a CR map if $F_*V \subset V_1$.

LEMMA 17. *A function* f *is a CR function with respect to the CR structure* $\{\omega, \omega_1, \ldots, \omega_n\}$ *if and only if*

(34) $df \in$ linear span $\{\omega, \omega_1, \ldots, \omega_n\}$.

LEMMA 18. *A map* $F: M \to M_1$ *is a CR map with respect to the CR structures* $\{\omega, \omega_1, \ldots, \omega_n\}$ *and* $\{\theta, \theta_1, \ldots, \theta_n\}$ *if and only if*

(35) $F^*(\theta_J) \in$ linear span $\{\omega, \omega_1, \ldots, \omega_n\}$ for $J = 0, 1, \ldots, n$.

A CR map is a diffeomorphism if and only if

(36) $$F^*(\theta \wedge \theta_1 \wedge \cdots \wedge \theta_n) \neq 0.$$

PROOFS. The statements (34) and (35) are just the duals of the recalled definitions. The proof that (36) is equivalent to

$$F^*(\theta \wedge \theta_1 \wedge \cdots \wedge \theta_n \wedge \overline{\theta}_1 \wedge \cdots \wedge \overline{\theta}_n) \neq 0$$

is similar to the proof of Lemma 10.

REMARK. These conditions can be rephrased as

$$df \in H^{0,1}(M)$$

and

$$F^*(H^{0,1}(M_1, q)) \subset H^{0,1}(M, p).$$

EXERCISE 14. Let $f_1, \ldots, f_{n+1}$ be CR functions on (M^{2n+1}, p) with $df_1 \wedge \cdots \wedge df_{n+1} \neq 0$ at p. Show that a function F is CR on (M, p) if and only if

$$dF \in \text{linear span}\{df_1, \ldots, df_{n+1}\}.$$

The CR structures in which we are interested have a natural orientation.

DEFINITION. An *orientation* on M^n is a nowhere vanishing section of the bundle of n-forms.

Since on a manifold of dimension n the bundle of n-forms is one-dimensional, there is at each point two choices for the orientation. A CR manifold is said to have a natural orientation if at each point one can make a choice using only CR information.

EXAMPLE. Either orientation on

$$M = \{(z, w): \text{Im } w = 0\}$$

is reversed by the restriction to M of the biholomorphism

$$z \to z, \qquad w \to -w.$$

Thus M does not have a natural CR orientation.

LEMMA 19. *There is a natural CR orientation on any strictly pseudoconvex* (M^3, V).

PROOF. Let $\{\omega, \omega_1\}$ give the CR structure near some point of M. Replace ω by $-\omega$ if necessary so as to obtain

$$d\omega \equiv ig\omega_1\overline{\omega}_1 \pmod{\omega}$$

with $g > 0$. Define the orientation near p by $\omega \wedge \omega_1 \wedge \overline{\omega}_1$. It is easy to see that any other choice $\{\theta, \theta_1\}$ with

$$d\theta \equiv ih\theta_1\overline{\theta}_1 (\mod \theta)$$

with $h > 0$, satisfies

$$\theta \wedge \theta_1 \wedge \overline{\theta}_1 = a\omega \wedge \omega_1 \wedge \overline{\omega}_1$$

with $a > 0$.

COROLLARY. *A CR diffeomorphism* $\phi: (M^3, V) \to (M_1^3, V_1)$ *of nonde-generate structures preserves the natural orientation.*

This means that if $\omega \wedge \omega_1 \wedge \overline{\omega}_1$ gives the orientation of (M, V) and $\theta \wedge \theta_1 \wedge \overline{\theta}_1$ gives the orientation of (M_1, V_1), then

$$\phi^*(\theta \wedge \theta_1 \wedge \overline{\theta}_1) = a\omega \wedge \omega_1 \wedge \overline{\omega}_1$$

with $a > 0$.

EXERCISE 15. Show that a CR manifold of signature (p, q, r) has a natural orientation provided $p \neq q$ and find an example of a CR manifold of signature (p, p, r) that admits an orientation reversing CR automorphism.

§6. Reinhardt hypersurfaces. We introduce a special class of hypersurfaces to illustrate some of the previous material. This class will be studied again in Chapter 8.

DEFINITION. A domain $\mathscr{D} \subset \mathbb{C}^N$ is called a *Reinhardt domain* if for each $Z = (z_1, \ldots, z_N) \in \mathscr{D}$ the set

$$\{W = (\lambda_1 z_1, \ldots, \lambda_N z_N) : |\lambda_j| \leq 1\}$$

is contained in $\mathscr{D}$. A smooth hypersurface $M^{2n+1} \subset \mathbb{C}^{n+1}$ is called a *Reinhardt hypersurface* if for each $Z = (z_1, \ldots, z_{n+1}) \in M$ the set

$$\{W = (\lambda_1 z_1, \ldots, \lambda_{n+1} z_{n+1}) : |\lambda_j| = 1\}$$

is contained in M.

It is clear that every Reinhardt domain contains the origin as an interior point; that the boundary of a Reinhardt domain is a Reinhardt hypersurface; and that no Reinhardt hypersurface contains the origin.

EXERCISE 16. Show that if near a point (z_0, w_0) a Reinhardt hypersurface has the form

$$M = \{(z, u + iv) \in \mathbb{C}^{n+1} : v = f(z, \overline{z}, u) \text{ and } |z - z_0| < \varepsilon\},$$

then M is a circle bundle over $\{z \in C^n : |z - z_0| < \varepsilon\}$ and so w_0 cannot be zero.

LEMMA 20. *Any Reinhardt hypersurface* M^{2n+1} *in* $\mathbb{C}^{n+1}$ *has a defining function of the form* $r = r(|z_1|^2, \ldots, |z_{n+1}|^2)$ *where* $r(x_1, \ldots, x_{n+1})$ *is a smooth function on* $\mathbb{R}^{n+1}$.

PROOF. It is enough to find local defining functions and then put them together in an obvious way. Given some point $p \in M$, we may relabel coordinates so that near p, M is given as

(37) $v = f(z, \overline{z}, u).$

Denote p by (z_0, ζ_0). We may not assume that $df(z_0, \overline{z}_0, u) = 0$.

The map $z \to e^{i\theta} z$ preserves M. So for θ small we also have

$$v = f(e^{i\theta} z, e^{-i\theta}\overline{z}, u).$$

Thus

$$f(z, \bar{z}, u) = g(|z|, u).$$

We set $h(s, u)$ equal to $g(\sqrt{s}, u)$. Since f is smooth, h is smooth for $s > 0$. We then extend h smoothly to negative values. We now have that M is given near (z_0, ζ_0) by

$$(38) \qquad\qquad v = h(|z|^2, u).$$

Thus it is given by this equation in a neighborhood of $(\lambda z_0, \zeta_0)$. By the Reinhardt property we know that, for fixed z, equation (38) defines a circle of some radius $(\sigma(|z|^2))^{1/2}$. Thus for the local defining function we may take

$$r(z, \zeta) = |\zeta|^2 - \sigma(|z|^2).$$

EXERCISE 17. Show that for any point over $(0, \zeta_0)$, H is given by the z-plane.

We now restrict ourselves to three-dimensional Reinhardt hypersurfaces, M^3. The family we are interested in are circle bundles over the deleted z-plane. We start with the coordinates

$$z = te^{i\theta_1} \text{ and } w = f(t)e^{i\theta_2}.$$

Note that f cannot be zero.

EXERCISE 18. Show that we may take for the Lewy operator

$$(39) \qquad\qquad L = \frac{\partial}{\partial t} + \frac{i}{t}\frac{\partial}{\partial \theta_1} + \frac{if'(t)}{f(t)}\frac{\partial}{\partial \theta_2}.$$

It will be useful to replace t by the new variable $s = \log t$. Then, for $g(s) = \log f(e^s)$, the equation of the hypersurface becomes

$$(40) \qquad\qquad \log |z_2| = g(\log |z_1|)$$

with local coordinates

$$(41) \qquad\qquad (s, \theta_1, \theta_2) \to (e^{s+i\theta_1}, e^{g(s)+i\theta_2}).$$

LEMMA 21.
(a) *We may take as the Lewy operator* $L = \frac{\partial}{\partial s} + i\frac{\partial}{\partial \theta_1} + ig'(s)\frac{\partial}{\partial \theta_2}$.
(b) *H is spanned by* $\frac{\partial}{\partial s}$ *and* $\frac{\partial}{\partial \theta_1} + g'(s)\frac{\partial}{\partial \theta_2}$.
(c) *The hypersurface is strictly pseudoconvex at* (s, θ_1, θ_2) *if and only if* $g''(s) \neq 0$.

For ω, we just need a real form that annihilates

$$L = \frac{\partial}{\partial s} + i\frac{\partial}{\partial \theta_1} + ig'(s)\frac{\partial}{\partial \theta_2}.$$

So we may take

$$\omega = d\theta_2 - g'(s)d\theta_1.$$

For ω_1 we take dz,

$$\omega_1 = d(e^{s+i\theta_1}) = e^{s+i\theta_1}(ds + id\theta_1).$$

Since

$$d\omega = -g''(s)ds\,d\theta_1$$

and

$$\omega_1\overline{\omega}_1 = 2e^{2s}id\theta_1\,ds\,,$$

we have

(42) $$d\omega = \frac{-i}{2}\frac{g''(s)}{e^{2s}}\omega_1\overline{\omega}_1 = \frac{-i}{2}\frac{g''(s)}{|z|^2}\omega_1\overline{\omega}_1\,.$$

Let us redefine ω to be

$$\omega = -\tfrac{1}{2}(d\theta_2 - g'(s)d\theta_1)\,,$$

and take $g(s)$ to be

$$g(s) = Bs + e^{2s}$$

where B is an arbitrary constant. Then

$$d\omega = i\omega_1\overline{\omega}_1\,.$$

Thus the induced CR structure on the hypersurface

$$\mathscr{R}_B = \{(z, w): |w| = |z|^B e^{|z|^2}\}$$

is defined by forms

$$\omega = -\frac{1}{2}(d\theta_2 - (B + 2e^{2s})d\theta_1)\,,$$

(43)   and

$$\omega_1 = e^{s+i\theta_1}(ds + id\theta_1)\,,$$

which satisfy

(44) $$d\omega = i\omega_1\overline{\omega}_1\,.$$

It is very easy to show that (44) implies each $\mathscr{R}_B$ is locally equivalent to Q. Several similar problems will be worked out in Chapter 5. The basic idea is to use the Frobenius Theorem to construct a local diffeomorphism Φ of neighborhoods in $\mathscr{R}_B$ and Q such that

$$\omega = \Phi^*(\tfrac{1}{2}du - \tfrac{i}{2}\overline{z}dz + \tfrac{i}{2}z d\overline{z})\,,$$
$$\omega_1 = \Phi^*(dz)\,.$$

Thus Φ is a CR diffeomorphism. Since $\mathscr{R}_B$ is real analytic, so is Φ. Thus Φ extends to a local biholomorphism. See Chapter 8 for an explicit map of $\mathscr{R}_B$ to Q and for a proof that, in general, $\mathscr{R}_{B_1}$ and $\mathscr{R}_{B_2}$ are not globally (or even "semiglobally") equivalent. Also note that $\mathscr{R}_B$ is smooth across $z_1 = 0$ only for $B = 0$.

There are more complicated Reinhardt hypersurfaces which are also locally equivalent to Q.

EXERCISE 19. Consider the hypersurface

$$|z_2| = |z_1|^B |z_1|^{C \log |z_1|}$$

for some constants B and C.

a) Show that

$$\omega = -\tfrac{1}{C}(d\theta_2 - (B + 2Cs)d\theta_1),$$
$$\omega_1 = ds + id\theta$$

gives the CR structure. (Hint: One way to do this is to show that $L \in \{\omega, \omega_1\}^{\perp}$ annihilates the functions $e^{s+i\theta_1}$ and $e^{sB+s^2C}e^{i\theta_2}$.)

b) Verify that

$$d\omega = i\omega_1 \overline{\omega}_1,$$
$$d\omega_1 = 0,$$

and conclude that this hypersurface is locally equivalent to Q.

CHAPTER 2

Some Automorphism Groups

The main results of this chapter are:

1. The assertion that the ball and the polydisc in $\mathbb{C}^2$ are not biholomorphically equivalent. This gives another demonstration, in a somewhat different sense, of the failure of the Riemann Mapping Theorem to extend to higher dimensions.

2. The computation of the automorphism group of the ball.

3. A geometric study of $SU(2, 1)$. This eight-dimensional group, and its five-dimensional isotropy subgroup, will be important in the work of Moser and of Cartan.

§1. Automorphism groups and the absence of a Riemann Mapping Theorem. We start the chapter by finding the automorphism group of the ball in $\mathbb{C}^1$, using homogeneous coordinates. Similar projective techniques will reappear throughout these notes. Let $\mathbb{CP}^1$ be the one-dimensional complex projective space. Points of $\mathbb{C}^2 - \{0\}$ provide homogeneous coordinates on $\mathbb{CP}^1$ by means of the equivalence relation $(\zeta_1, \zeta_2) \sim (\xi_1, \xi_2)$ if $(\zeta_1, \zeta_2) = \lambda(\xi_1, \xi_2)$ for some complex number λ. We let $[\zeta_1, \zeta_2]$ denote the set of points in $\mathbb{C}^2 - \{0\}$ equivalent to (ζ_1, ζ_2) and also the point in $\mathbb{CP}^1$ given by the complex line determined by (ζ_1, ζ_2) and the origin. We seek a family of automorphisms of

$$B = \{z \in \mathbb{C}^1 : |z| < 1\}$$

by considering the projective linear maps. Recall that a linear map on $\mathbb{C}^2$ provides a well-defined map of $\mathbb{CP}^1$ to itself: The linear map with matrix

$$A = \begin{pmatrix} a_{00} & a_{01} \\ a_{10} & a_{11} \end{pmatrix}$$

gives $[\zeta_0, \zeta_1] \rightarrow [a_{00}\zeta_0 + a_{01}\zeta_1, a_{10}\zeta_0 + a_{11}\zeta_1]$. This in turn gives a well-defined meromorphic map of $\mathbb{C}^1$:

$$z = \frac{\zeta_1}{\zeta_0} \rightarrow \frac{a_{10}\zeta_0 + a_{11}\zeta_1}{a_{00}\zeta_0 + a_{01}\zeta_1} = \frac{a_{10} + a_{11}z}{a_{00} + a_{01}z}.$$

We obtain the same map if we replace A by λA, for any nonzero scalar λ. Note that the unit ball in $\mathbb{C}^1$ corresponds to the set $|\zeta_1|^2 - |\zeta_0|^2 < 0$ in $\mathbb{CP}^1$. For any ξ and η in $\mathbb{C}^2$, let $\langle \xi, \eta \rangle = \xi_1 \bar{\eta}_1 + \xi_2 \bar{\eta}_2$. Thus the unit ball corresponds to $\zeta \in \mathbb{C}^2$ with $\langle C\zeta, \zeta \rangle < 0$,

$$C = \begin{pmatrix} -1 & 0 \\ 0 & 1 \end{pmatrix}.$$

LEMMA 1. *A linear map of $\mathbb{C}^2$ which preserves the bilinear form corresponding to C gives an automorphism of B.*

Let us determine such automorphisms explicitly. Assume

$$A = \begin{pmatrix} \alpha & \beta \\ \gamma & \delta \end{pmatrix}$$

preserves the bilinear form. This means $A^* C A = C$, where

$$A^* = \begin{pmatrix} \bar{\alpha} & \bar{\gamma} \\ \bar{\beta} & \bar{\delta} \end{pmatrix}$$

is the (Hermitian) adjoint of A. These matrices form the group $\mathrm{U}(1, 1)$. Note $|\det A| = 1$ and, since we are interested in the mapping on $\mathbb{C}^1$, we replace A by λA to obtain $\det A = 1$. Thus

$$A^{*-1} = \begin{pmatrix} \bar{\delta} & -\bar{\gamma} \\ -\bar{\beta} & \bar{\alpha} \end{pmatrix},$$

and if we rewrite $A^* C A = C$ as $C A = (A^*)^{-1} C$ we see that

$$\begin{pmatrix} \alpha & \beta \\ -\gamma & -\delta \end{pmatrix} = \begin{pmatrix} \bar{\delta} & \bar{\gamma} \\ -\bar{\beta} & -\bar{\alpha} \end{pmatrix}.$$

From this we conclude $\alpha = \bar{\delta}$, $\gamma = \bar{\beta}$, $|\alpha|^2 - |\beta|^2 = 1$. Thus the automorphism is $z \to \frac{\bar{\alpha} z + \bar{\beta}}{\beta z + \alpha}$ where $|\alpha|^2 - |\beta|^2 = 1$. Let $a = \frac{\bar{\beta}}{\bar{\alpha}}$ and $e^{i\theta} = \frac{\bar{\alpha}}{\alpha}$. Note $|a| < 1$. This gives us a three-parameter family of automorphisms of the ball $z \to e^{i\theta}(\frac{z+a}{1+\bar{a}z})$, where $\theta \in \mathbb{R}$ and $a \in B$. Note that any point of B can be taken to the origin. These automorphisms form a group "essentially" isomorphic to the subgroup of $\mathrm{GL}(2, \mathbb{C})$ consisting of matrices A with $A^* C A = C$ and $\det A = 1$. This group is denoted $\mathrm{SU}(1, 1)$. The qualification "essentially" is added only because the action of $\mathrm{SU}(1, 1)$ on $\mathbb{CP}^1$ is not effective, since all points are fixed by both the identity and the negative of the identity.

THEOREM 1. *Every automorphism of the unit ball in $\mathbb{C}^1$ is of the form $\phi(z) = e^{i\theta} \frac{z+a}{1+\bar{a}z}$ where θ is real and a is a point of this ball.*

We use two lemmas in the proof.

SCHWARZ LEMMA. *If f is a holomorphic map of B into itself which preserves the origin then $|f(z)| \leq |z|$ for all $z \in B$ and $|f'(0)| \leq 1$. If $|f'(0)| = 1$ or if $|f(z)| = |z|$ for some nonzero z then $f(z) = \lambda z$ for some λ with $|\lambda| = 1$.*

PROOF. Let $h(z) = f(z)/z$. So h is holomorphic on B and $|h(z)| \leq (1 - \varepsilon)^{-1}$ on $\{|z| = 1 - \varepsilon\}$. By the maximum modulus principle $|h(z)| \leq (1 - \varepsilon)^{-1}$ on $\{|z| \leq 1 - \varepsilon\}$ and so $|f(z)| \leq (1 - \varepsilon)^{-1}|z|$ on $\{|z| \leq 1 - \varepsilon\}$ for all small ε. Thus $|h(z)| \leq 1$ and $|f(z)| \leq |z|$ on B. If $|f(z_0)| = |z_0|$ for some $z_0 \neq 0$ then $|h(z_0)| = 1$. But this implies $h(z)$ is a constant, say λ. Thus $f(z) = \lambda z$. Finally, $f'(0) = \lim_{z \to 0} \frac{f(z)}{z}$ so $|f'(0)| \leq 1$; and if $|f'(0)| = 1$, then $|h(0)| = 1$ and again $f(z) = \lambda z$.

LEMMA 2. *The only automorphisms of B which preserve the origin are $f(z) = \lambda z$, $|\lambda| = 1$.*

PROOF. We apply the Schwarz lemma to f and to $g = f^{-1}$. We have $|f'(0)| \leq 1$ and $|g'(0)| \leq 1$ and so $|f'(0)| = 1$. It follows that $f(z) = \lambda z$.

Now let $F: B \to B$ be any automorphism. Let $h: B \to B$ be one of the automorphisms of $SU(1, 1)$ which is chosen to take $F(0)$ to 0. So $h \circ F$ is an automorphism of B which preserves the origin. Thus by this lemma, $h \circ F(z) = \lambda z$ and so $F(z) = h^{-1}(\lambda z)$. F is a product of elements of $SU(1, 1)$ and therefore is itself in $SU(1, 1)$. This completes the proof of the theorem.

We now consider the unit ball in $\mathbb{C}^2$. So henceforth we use B for this ball,

$$B = \{(z_1, z_2) \in \mathbb{C}^2 : |z_1|^2 + |z_2|^2 < 1\}.$$

We also consider the polydisc

$$P = \{(z_1, z_2) \in \mathbb{C}^2 : |z_1| < 1 \text{ and } |z_2| < 1\}.$$

In general, let $\text{Aut}(\Omega)$ denote the group of biholomorphic mappings into itself of some open set Ω in $\mathbb{C}^2$. We will determine $\text{Aut}(P)$ and $\text{Aut}(B)$ in the same way that we determined the automorphisms of the ball in $\mathbb{C}^1$. That is, we explicitly find some group of automorphisms and then show there can be no others. In place of the Schwarz lemma we use the following result.

H. CARTAN LEMMA. *Let $f: D \to D$ be a holomorphic mapping of some bounded domain D into itself, $D \subset \mathbb{C}^n$, $n \geq 1$. Assume f leaves some point fixed and $df = \text{Id}$ at that point. Then f is the identity map.*

PROOF. We may take the fixed point to be the origin. Find r and R so that

$$\{z \in \mathbb{C}^n : |z| < r\} \subset D \subset \{z \in \mathbb{C}^n : |z| < R\}.$$

For a holomorphic function

$$F(z) = F(z_1, \ldots, z_n) = \sum_{|\alpha| = 0}^{\infty} a_\alpha z^\alpha,$$

we have Cauchy's estimate

$$|a_\alpha| \le r^{-|\alpha|} \max_{|z_j|=r} |F|.$$

We are starting with a mapping f of the form

$$f(z_1, \ldots, z_n) = \begin{pmatrix} z_1 \\ \vdots \\ z_n \end{pmatrix} + \begin{pmatrix} P_{N_1}(z_1, \ldots, z_n) \\ \vdots \\ P_{N_n}(z_1, \ldots, z_n) \end{pmatrix} + \cdots$$

where each P_{N_j} is a polynomial of degree N_j. That is,

$$f(z) = z + P_N(z) + \cdots$$

where $P_N(z)$ is a vector of polynomials of degree $N = \min_j\{N_j\} \ge 2$ and either at least one component of $P_N(z)$ is not identically zero or else $f(z) \equiv z$. If we compose f with itself k times we obtain

$$f^{(k)}(z) = z + k P_N(z) + \cdots .$$

Let $F(z)$ be any component of $f^{(k)}(z)$. Since $f^{(k)}$ is also a map of D to itself we see that $|F(z)| < R$ for each $z \in D$. Thus each coefficient in the Taylor expansion for F is bounded by $R/r^{|\alpha|}$. But this is independent of k. Thus all components of $P_N(z)$ must be identically zero which means that f is the identity. This proves the lemma.

It might be expected that we use this lemma by starting with a known group of automorphisms, considering an arbitrary automorphism, and reducing to the situation of this lemma by forming a composition. This is not quite so. An automorphism group can never be transitive on the one-jet level since automorphisms are isometric with respect to the Bergman metric. However, for a wide class of domains, including P and B, we can carry out this reduction by a different argument.

DEFINITION. $D \subset \mathbb{C}^n$ is *circular* if $0 \in D$ and for each $z \in D$ one also has $e^{i\theta} z \in D$ for all real θ.

REMARK. A Reinhardt domain is a special circular domain where θ may vary on the different components of z.

LEMMA 3 (H. CARTAN). *Let $f : D \to D$ be a holomorphic mapping of some bounded circular domain into itself leaving the origin fixed. Then f is a linear map.*

PROOF. Let $k_\theta(z) = e^{i\theta} z$ and $g(z) = k_\theta^{-1} \circ f^{-1} \circ k_\theta \circ f$. So $g : D \to D$ and $g(0) = 0$. Further $dg|_0 = \mathrm{id}$. Thus, by the previous lemma, $g(z) = z$ and so

$$e^{i\theta} f(z) = f(e^{i\theta} z).$$

It follows that f is linear.

It is easy, using these two lemmas of Cartan, to determine all the automorphisms of the polydisc. Since P is the product of unit balls in $\mathbb{C}^1$

we can immediately write down two types of automorphisms of P. For $\theta_j \in \mathbb{R}$, $\alpha_j \in \mathbb{C}^1$ with $|\alpha_j| < 1$, for $j = 1$ and 2, set

$$\phi_{\theta,\alpha}(z_1, z_2) = \left(e^{i\theta_1} \frac{z_1 + \alpha_1}{1 + \bar{\alpha}_1 z_1}, \; e^{i\theta_2} \frac{z_2 + \alpha_2}{1 + \bar{\alpha}_2 z_2}\right),$$

and set $\sigma(z_1, z_2) = (z_2, z_1)$.

THEOREM 2. *Each automorphism of the polydisc is given either by* $\phi_{\theta,\alpha}$ *or* $\phi_{\theta,\alpha} \circ \sigma$.

PROOF. Let $F: P \to P$ be an automorphism. Choose some $h = \phi_{\theta,\alpha}$ such that $h \circ F(0) = 0$. Thus by Lemma 3, $h \circ F$ is a linear map of P to itself. So we now determine these linear maps.

LEMMA 4. *Each linear map of the polydisc to itself is given either by* $\phi(z_1, z_2) = (e^{i\theta_1} z_1, \; e^{i\theta_2} z_2)$ *or by* $\phi \circ \sigma$.

Before proving this lemma we show that the theorem follows. For $h \circ F$ is linear, so either $F = h^{-1}(e^{i\theta_1} z_1, \; e^{i\theta_2} z_2)$ or $F = h^{-1}(e^{i\theta_1} z_2, \; e^{i\theta_2} z_1)$. Thus F is either given by some $\phi_{\theta,\alpha}$ or by $\phi_{\theta,\alpha} \circ \sigma$.

PROOF OF LEMMA. We start with some linear map for P to itself and take as its matrix

$$A = \begin{pmatrix} a_{11} & a_{12} \\ a_{21} & a_{22} \end{pmatrix}.$$

Since A maps $(1, 0)$ to a boundary point, we have
  (i) $|a_{11}| \leq 1$, $|a_{21}| \leq 1$ and at least one is an equality.
Similarly A maps $(0, 1)$ to a boundary point and we have
  (ii) $|a_{12}| \leq 1$, $|a_{22}| \leq 1$ and at least one is an equality.
Thus $(\bar{a}_{11}, \bar{a}_{12})$ is in the closure of P. The same result must be true for $A(\bar{a}_{11}, \bar{a}_{12})$. Thus
  (iii) $|a_{11}|^2 + |a_{12}|^2 \leq 1$.
Similarly,
  (iv) $|a_{21}|^2 + |a_{22}|^2 \leq 1$.
Now if $|a_{11}| = 1$ then from (iii), (ii), (iv), we have

$$A = \begin{pmatrix} e^{i\theta_1} & 0 \\ 0 & e^{i\theta_2} \end{pmatrix},$$

while if $|a_{11}| < 1$ then from (i), (iv), (ii), (iii), we have

$$A = \begin{pmatrix} 0 & e^{i\theta_1} \\ e^{i\theta_2} & 0 \end{pmatrix}.$$

This finishes the proof of the lemma and so also of the theorem.

Even before finding the automorphism group of the unit ball we may show that $\mathrm{Aut}(B)$ is not isomorphic to $\mathrm{Aut}(P)$ and so B and P are not biholomorphically equivalent.

THEOREM 3. (USUALLY CREDITED TO POINCARÉ). *There is no biholomorphic map from the ball B to the polydisc P.*

PROOF. Let

$$\mathcal{F}(\Omega) = \{f \in \operatorname{Aut}(\Omega): f(0) = 0 \text{ and there exists}$$
$$f_t \in \operatorname{Aut}(\Omega) \text{ with } f_t(0) = 0, \ f_0 = f, \ f_1 = \text{identity}\}.$$

Here of course we require that f_t varies "continuously" in t. This is no problem since, as we shall see, $\operatorname{Aut}(B)$ is a closed subgroup of $\operatorname{GL}(3, C)$.

Now $\mathcal{F}(P)$ consists of the maps $(z_1, z_2) \to (e^{i\theta_1} z_1, e^{i\theta_2} z_2)$ and is an abelian group. But

$$F_{\theta,\sigma} = \begin{pmatrix} \cos\theta & -e^{-i\sigma}\sin\theta \\ e^{i\sigma}\sin\theta & \cos\theta \end{pmatrix}$$

is an element of the identity component of $U(2)$. That is, it preserves the Euclidean metric in C^2 and so maps B to itself. It clearly also preserves the origin. Thus each $f_{\theta,\sigma} \in \mathcal{F}(B)$. But $f_{\frac{\pi}{4},0} \circ f_{\frac{\pi}{4},\sigma} \neq f_{\frac{\pi}{4},\sigma} \circ f_{\frac{\pi}{4},0}$. So $\mathcal{F}(B)$ is not abelian. It is easy to see that this implies that B and P are not biholomorphically equivalent. For let

$$\Phi: B \to P$$

be a biholomorphism and let

$$\Psi: P \to P$$

be any automorphism which takes $\Phi(0)$ to 0. Then

$$F: \mathcal{F}(B) \to \mathcal{F}(P)$$

given by

$$F = \Psi\Phi\mu(\Psi\Phi)^{-1}, \qquad \mu \in \mathcal{F}(B),$$

is an isomorphism. But this is impossible since the second group is abelian and the first is not.

We now compute $\operatorname{Aut}(B)$ in the same way as we did for the ball in C^1. In terms of the homogeneous coordinates $z_1 = \frac{\zeta_1}{\zeta_0}$, $z_2 = \frac{\zeta_2}{\zeta_0}$ we first seek the linear maps of C^2 which preserve the quadratic form $\langle \zeta, \zeta \rangle = -|\zeta_0|^2 + |\zeta_1|^2 + |\zeta_2|^2$. The group of such linear maps is denoted by $U(2, 1)$. We shall again use the group $U(2)$, the unitary group on C^2, which consists of linear maps which preserve the Euclidean form

$$\langle z, z \rangle = |z_1|^2 + |z_2|^2.$$

Let

$$A = \begin{pmatrix} a_{00} & a_{01} & a_{02} \\ a_{10} & a_{11} & a_{12} \\ a_{20} & a_{21} & a_{22} \end{pmatrix}$$

and

$$C = \begin{pmatrix} -1 & 0 & 0 \\ 0 & 1 & 0 \\ 0 & 0 & 1 \end{pmatrix}.$$

Then $U(2, 1) = \{A: A^*CA = C\}$. Similarly, if instead A denotes an $n \times n$ matrix, then $U(n) = \{A: A^*A = \mathrm{Id}\}$. Note that the dimensions (over $\mathbb{R}$) of $U(2)$ and $U(2, 1)$ are four and nine, respectively, and that in its action over $\mathbb{C}^2$, $U(2, 1)$ has dimension eight.

The following lemma is easily proved by counting unknowns and equations.

LEMMA 5. 1. $U(2, 1)$ is transitive on $\mathbb{C}^3$ in the sense that if $\langle \zeta, \bar{\zeta} \rangle = \langle \eta, \bar{\eta} \rangle$ then there is some $A \in U(2, 1)$ such that $A\zeta = \eta$.
2. The isotropy subgroup of any nonzero point is of dimension 4.

Note that the action of $U(2, 1)$ on $\mathbb{C}^2$ is given by

$$(z_1, z_2) \to \left(\frac{a_{10} + a_{11}z_1 + a_{12}z_2}{a_{00} + a_{01}z_1 + a_{02}z_2}, \frac{a_{20} + a_{21}z_1 + a_{22}z_2}{a_{00} + a_{01}z_1 + a_{02}z_2} \right).$$

This action is not effective since $(\zeta_0, \zeta_1, \zeta_2) \to e^{i\theta}(\zeta_0, \zeta_1, \zeta_2)$ is a nontrivial element of $U(2, 1)$ which gives the identity map on $\mathbb{C}^2$.

We may restate the above lemma with a bit more detail in terms of the action of $U(2, 1)$ on the ball $B \subset \mathbb{C}^2$.

LEMMA 6. 1. $U(2, 1)$ is transitive on B.
2. The isotropy subgroup of any point is isomorphic to the action of

$$\left(\begin{array}{c|cc} a & 0 & 0 \\ \hline 0 & & \\ & & M \\ 0 & & \end{array} \right)$$

with $a \in U(1)$ and $M \in U(2)$. (This action gives the isotropy subgroup of the origin.)

In particular, the isotropy subgroup is of dimension 5 and B can be identified with the homogeneous space

$$U(2, 1)/U(1) \times U(2).$$

Now let us show that $U(2, 1)$ gives all the automorphisms of B.

THEOREM 4. Let $f: B \to B$ be biholomorphic. Then

$$f(z_1, z_2) = \left(\frac{a_{10} + a_{11}z_1 + a_{12}z_2}{a_{00} + a_{01}z_1 + a_{02}z_2}, \frac{a_{20} + a_{21}z_1 + a_{22}z_2}{a_{00} + a_{01}z_1 + a_{02}z_2} \right)$$

where the matrix A of the coefficients is in $U(2, 1)$.

PROOF. We have stated that $U(2, 1)$ acts transitively on B. So there is some $G \in U(2, 1)$ for which, in the action on B, $G \circ f(0) = 0$, and, of course, $G \circ f: B \to B$. Thus by Lemma 3, $G \circ f$ is linear. But U(2) gives all the linear maps of B to itself. Thus $G \circ f \in U(2) \subset U(2, 1)$ and hence f is given by the action of some $A \in U(2, 1)$ on $\mathbb{C}^2$.

Note that Theorem 1 asserts that every automorphism of the ball in $\mathbb{C}^1$ is given by the composition of a unitary transformation (multiplication by $e^{i\theta}$) and a particular automorphism $(\phi_a = \frac{z+a}{1+\bar{a}z})$ which takes 0 to some given point of the ball. This is true also in higher dimensions. We use the notation of Rudin [**Ru**].

EXERCISE 1. Let $a \in B$, $a \neq 0$. Let P_a denote the orthogonal projection of $\mathbb{C}^2$ onto the complex line generated by a and let $Q_a = I - P_a$ be the projection onto the orthogonal complement of this line. Put $S_a = (1 - |a|^2)^{\frac{1}{2}}$ and set

$$\phi_a = \frac{a - P_a Z - S_a Q_a Z}{1 - \langle z, a \rangle}.$$

(a) Show that ϕ_a is an automorphism of B.
(b) Show that any other automorphism Ψ of B has the form $\Psi = U\phi_a$ for some $a \in B$ and some $U \in U(2)$. (Hint: The solution is in Chapter 2 of [**Ru**].)

§2. The group $SU(2, 1)$.

We want to describe more explicitly the automorphisms of B. We start by looking at the elements of $U(2, 1)$. For later work it is convenient to work with an unbounded region which is biholomorphically equivalent to B.

Let

$$H = \{(z, w): \operatorname{Im} w > |z|^2\} \quad \text{and} \quad Q = \{(z, w): \operatorname{Im} w = |z|^2\}.$$

LEMMA 7. B and H are biholomorphically equivalent.

PROOF. The map $z = \frac{z_1}{z_2 - i}$, $w = \frac{1 - iz_2}{z_2 - i}$ establishes the equivalence. Note that the map of the boundaries takes $(0, i)$ to infinity and $(0, -i)$ to the origin.

REMARKS. 1. This map extends to a CR map of the boundaries, $\phi: S^3 \to Q$, where S^3 is the sphere and Q is the hyperquadric. We say that S^3 and Q are CR diffeomorphic (although we must allow some point of S to map to ∞).

2. The map in the lemma is projective, so it takes complex lines to complex lines. Thus to study, for instance, the intersection of S^3 with complex lines we may work instead with Q. We do this in Chapter 8.

Let us write the boundary Q in homogeneous coordinates. So $z = \frac{\zeta_1}{\zeta_0}$ and $w = \frac{\zeta_2}{\zeta_0}$. From $\operatorname{Im} w = |z|^2$ we derive

$$|\zeta_1|^2 + \frac{i}{2}(\zeta_2\bar{\zeta}_0 - \bar{\zeta}_2\zeta_0) = 0.$$

We denote the left-hand side by $\langle \zeta, \zeta \rangle$ and also write

$$\langle \zeta, \zeta \rangle = \zeta \cdot C\bar{\zeta}$$

where C is the matrix

$$\begin{pmatrix} 0 & 0 & -i/2 \\ 0 & 1 & 0 \\ i/2 & 0 & 0 \end{pmatrix}.$$

The set of matrices $A \in \mathrm{GL}(3, \mathbb{C})$ satisfying $A^*CA = C$ is another representation of $\mathrm{U}(2, 1)$. Writing out these conditions on A we have the following six equations. Again we label the elements of A by $a_{00}, a_{01}, \ldots, a_{22}$.

$$\frac{i}{2}\bar{a}_{00}a_{20} + |a_{10}|^2 - \frac{i}{2}\bar{a}_{20}a_{00} = 0,$$

$$\frac{i}{2}\bar{a}_{00}a_{21} + \bar{a}_{10}a_{11} - \frac{i}{2}\bar{a}_{20}a_{01} = 0,$$

$$\frac{i}{2}\bar{a}_{00}a_{22} + \bar{a}_{10}a_{11} - \frac{i}{2}\bar{a}_{20}a_{01} = \frac{i}{2},$$

$$\frac{i}{2}\bar{a}_{01}a_{21} + |a_{11}|^2 - \frac{i}{2}a_{01}\bar{a}_{21} = 1,$$

$$\frac{i}{2}\bar{a}_{01}a_{22} + \bar{a}_{11}a_{12} - \frac{i}{2}\bar{a}_{21}a_{02} = 0,$$

$$\frac{i}{2}\bar{a}_{02}a_{22} + |a_{12}|^2 - \frac{i}{2}\bar{a}_{22}a_{02} = 0.$$

We want an explicit representation of the isotropy subgroup of the origin in $\mathbb{C}^2$. So we assume

$$A\begin{pmatrix} 1 \\ 0 \\ 0 \end{pmatrix} = \begin{pmatrix} \lambda \\ 0 \\ 0 \end{pmatrix}$$

for some $\lambda \neq 0$. That is, we set $a_{00} = \lambda$, $a_{10} = 0$, $a_{20} = 0$. Then the first equation is satisfied and the other equations give us $a_{21} = 0$, $a_{22} = \bar{\lambda}^{-1}$, $|a_{11}| = 1$, $\bar{a}_{01} = 2i\bar{\lambda}\bar{a}_{11}a_{12}$, $\mathrm{Re}(ia_{02}\lambda^{-1}) = |a_{12}|^2$. Set $a_{11} = \mu$, $a_{12} = \mu a$. Thus $a_{01} = -2i\lambda\bar{a}$ and $a_{02} = -i\lambda|a|^2 + r\lambda$, r an arbitrary real number. Thus

$$g = \begin{pmatrix} \lambda & -i\lambda\bar{a} & -i\lambda|a|^2 + r\lambda \\ 0 & \mu & \mu a \\ 0 & 0 & 1/\bar{\lambda} \end{pmatrix}$$

with $\lambda \in \mathbb{C}$, $\lambda \neq 0$, $a \in \mathbb{C}$, $|\mu| = 1$, and $r \in \mathbb{R}$, is the general element of the isotropy subgroup.

Since we are interested in the action on $\mathbb{C}^2$ we may divide every element by λ. This makes the action on $\mathbb{C}^2$ effective; only the identity leaves every point fixed. Setting $\rho = \frac{1}{|\lambda|^2}$ and $C = \frac{\mu}{\lambda}$ we obtain

(1)
$$g = \begin{pmatrix} 1 & -2i\bar{a} & r - i|a|^2 \\ 0 & C & Ca \\ 0 & 0 & \rho \end{pmatrix}$$

where $C \in \mathbb{C}$, $C \neq 0$, $a \in \mathbb{C}$, $r \in \mathbb{R}$, and $\rho = |C|^2$.

This is the isotropy subgroup of the origin in the group of all biholomorphic mappings on $\mathbb{C}^2$ which take Q to itself. Here we speak somewhat loosely; the mappings are permitted to have a singularity corresponding to the point at infinity on Q. This isotropy group has dimension five.

EXERCISE 2. Let $\phi(x, w) = (f(z, w), g(z, w))$ be an automorphism of Q. Show that the conditions:

$$\phi(0) = 0, \qquad d\phi(0) = \mathrm{id}, \qquad \mathrm{Re}\, \frac{\partial^2 g}{\partial w^2}(0) = 0$$

imply that ϕ is the identity.

Instead of dividing by λ, we can restrict ourselves to elements of $U(2, 1)$ which have determinant equal to one. This gives the group $SU(2, 1)$. The action is still not effective: For $g \in SU(2, 1)$ and $\lambda^3 = 1$, $\lambda \neq 1$, we have that g and λg are different as elements of $SU(2, 1)$ but the same as maps on $\mathbb{C}^2$.

We shall use the following notation:

$$S^3 = \{(z_1, z_2): |z_1|^2 + |z_2|^2 = 1\},$$

$$Q = \left\{(z_1, z_2): \mathrm{Im}\, z_2 = \frac{1}{2}|z_1|^2\right\},$$

$$\tilde{S} = \{(z_0, z_1, z_2): |z_1|^2 + |z_2|^2 - |z_0|^2 = 0\},$$

$$\tilde{Q} = \{(z_0, z_1, z_2): -iz_0\bar{z}_2 + iz_2\bar{z}_0 + |z_1|^2 = 0\},$$

$$F(Z, W) = -iz_0\bar{w}_2 + iz_2\bar{w}_0 + z_1\bar{w}_1,$$

$$F(Z) = F(Z, Z),$$

$$Z \cdot W = z_0 w_0 + z_1 w_1 + z_2 w_2.$$

Note that this Q differs by a factor of two from the one used previously. We will comment further on this in a moment.

We also let S and Q denote the corresponding compact objects in $\mathbb{CP}^2$. We do this by identifying $\mathbb{C}^2$ with $\{[1, z_1, z_2]\} \in \mathbb{CP}^2$. Thus

$$S = \{[z_0, z_1, z_2]: |z_1|^2 + |z_2|^2 - |z_0|^2 = 1\} = [\tilde{S}],$$

$$Q = \{[z_0, z_1, z_2]: F(z) = 0\} = [\tilde{Q}].$$

We will be working with $\mathbb{C}^2, \mathbb{C}^3$, and $\mathbb{CP}^2$. There is a linear relation among them: a complex 2-plane in $\mathbb{C}^3$ projects to some $\mathbb{CP}^1$ in $\mathbb{CP}^2$ which in turn projects to a complex line in $\mathbb{C}^2$. It is obvious, but well worth bearing in mind, that a real linear object in $\mathbb{C}^3$ in general does not project to a linear object in $\mathbb{C}^2$. We will especially need to remember this for Chapter 9.

EXAMPLE. Let P_1 be the real 3-plane in $\mathbb{C}^3$, which is spanned by $\{(1, 0, 0), (0, 1, 0), (0, 0, 1)\}$. If π is the composite projection of $\mathbb{C}^3$ into $\mathbb{C}^2$ given by $\pi(z_1, z_2, z_3) = (z_2/z_1, z_3/z_1)$ then πP_1 is the 2-plane $\mathrm{Im}\, z = 0, \mathrm{Im}\, w = 0$. Let P_2 be the real plane in $\mathbb{C}^3$ that is spanned by

$\{(\mu_1, 0, 0), (0, \mu_2, 0), (0, 0, \mu_3)\}$ where μ_j is some complex number. The projections of P_1 and P_2 both contain the point $(0, 0)$ and the points at infinity corresponding to the homogeneous coordinates $[0, 1, 0]$ and $[0, 0, 1]$. So if πP_2 is a plane, it must coincide with πP_1. But πP_2 contains $\pi[\mu_1, \mu_2, \mu_3] = (\mu_2/\mu_1, \mu_3/\mu_1)$ and this point is in P_1 if and only if μ_2/μ_1 and μ_3/μ_1 are real.

Here is a simple consequence. Let p, q, r be points of $\mathbb{C}^2$ and let A_1, A_2, A_3 be vectors in $\mathbb{C}^3$ with $[A_1] = p$, $[A_2] = q$, and $[A_3] = r$.

LEMMA 8. *The vectors A_1, A_2, A_3 are linearly independent over $\mathbb{C}$ if and only if the points p, q, r do not all lie on a complex line.*

PROOF. If the vectors are not independent then they span a complex 2-plane which projects to a line. Hence the points are collinear. Conversely, if the points are collinear then there is some complex 2-plane which projects to this line and hence must contain the vectors.

There is a nice geometric interpretation of $SU(2, 1)$, due to Cartan [Ca2], which will be important to us at various times. In order to facilitate reading this paper, we are using (more or less) Cartan's notation. In particular, we use $\widetilde{Q}$ defined by $F(Z) = 0$. Previously we used as the defining function

$$\langle \zeta, \zeta \rangle = -\frac{1}{2}i\zeta_0\overline{\zeta}_2 + \frac{1}{2}i\zeta_2\overline{\zeta}_0 + \zeta_1\overline{\zeta}_1.$$

The map

$$(z_0, z_1, z_2) \to (\zeta_0, \zeta_1, \zeta_2) \equiv (z_0, z_1/\sqrt{2}, z_2)$$

takes the hyperquadratic $F(Z) = 0$ to the hyperquadratic $\langle \zeta, \zeta \rangle = 0$. In terms of nonhomogeneous coordinates

$$z = z_1/z_0, \qquad w = z_2/z_0$$

and

$$x = \zeta_1/\zeta_0, \qquad y = \zeta_2/\zeta_0,$$

the map

(2) $$(z, w) \to (z/\sqrt{2}, w) \equiv (x, y)$$

takes the hyperquadratic

(3) $$\operatorname{Im} w = (1/2)|z|^2$$

to the hyperquadratic

(4) $$\operatorname{Im} y = |x|^2,$$

so the formulas for $SU(2, 1)$ with respect to $F(Z, Z)$ and with respect to $\langle \zeta, \zeta \rangle$, i.e., the formulas for the automorphism groups of (3) and (4), will differ by various factors of 2.

We now consider the representation of $SU(2, 1)$ which preserves F:

$$SU(2, 1) = \{A \in GL(3, \mathbb{C}) | F(AZ, AW) = F(Z, W) \text{ for all } Z \text{ and } W \text{ in } \mathbb{C}^3, \det A = 1\}.$$

Let

$$C = \begin{pmatrix} 0 & 0 & -i \\ 0 & 1 & 0 \\ i & 0 & 0 \end{pmatrix}.$$

So $F(Z, W) = Z \cdot C\overline{W}$. Note that

(5) $\overline{F(Z, W)} = F(W, Z),$

and also that

(6) $A \in \mathrm{SU}(2, 1)$ if and $only$ if $A^{\mathrm{T}} C \overline{A} = C.$

Let $A = (A_0, A_1, A_2)$ where A_j is a column vector in $\mathbb{C}^3$. From (6) we have:

LEMMA 9. $A \in \mathrm{SU}(2, 1)$ *if and only if* $\det(A_0, A_1, A_2) = 1$ *and each* $F(A_j, A_k) = 0$ *except for*

$$F(A_1, A_1) = 1, \qquad F(A_0, A_2) = -i, \qquad F(A_2, A_0) = i.$$

This is a total of ten real equations and hence $\dim \mathrm{SU}(2, 1) = 18 - 10 = 8$. This of course is consistent with the fact that $\dim \mathrm{U}(2, 1) = 9$.

Our geometric interpretation of $\mathrm{SU}(2, 1)$ will be to identify columns of $A \in \mathrm{SU}(2, 1)$ with certain points in $\mathbb{C}^2$ or $\mathbb{C}^3$. So for each $j = 0, 1, 2,$ let A_j define a point in $\mathbb{C}^3$ and let $[A_j]$ define the corresponding point in $\mathbb{C}\mathbb{P}^2$. From the lemma,

$$F(A_0, A_0) = 0, \qquad F(A_2, A_2) = 0,$$

and thus $[A_0]$ and $[A_2]$ are points in Q. Two other equations are

$$F(A_0, A_1) = 0 \quad \text{and} \quad F(A_2, A_1) = 0.$$

We claim that these two are equivalent to the statement:

$[A_1]$ *is the unique point of intersection of the complex lines tangent to* Q *at the points* $[A_0]$ *and* $[A_2]$.

Let us formalize this.

LEMMA 10. *Let* $p \in Q \subset \mathbb{C}\mathbb{P}^2$ *and let* L *be the complex line (i.e., copy of* $\mathbb{C}\mathbb{P}^1$ *) which is tangent to* Q *at* p. *Let* A *and* B *be vectors in* $\mathbb{C}^3$ *with* $[A] = p$. *Then the point* $[B]$ *is on* L *if and only if*

(7) $F(A, B) = 0.$

PROOF. The complex line through the points A and B is tangent to $\tilde{Q} \subset \mathbb{C}^3$ at A if and only if

$$dF(A + \zeta B) = 0 \quad \text{at} \quad \zeta = 0,$$

and this is equivalent to (7). Now if $[B]$ is on L then the line $A + \zeta B$ is tangent to $\tilde{Q}$ and so (7) holds. Conversely if (7) holds then $A + \zeta B$ is tangent. Either this line projects to L or it, and in particular B, projects to the point $[A]$. In either case, $[B] \in L$.

COROLLARY. *Let p and q be distinct points of $Q \subset \mathbb{CP}^2$ and let L_1 and L_2 be the corresponding complex tangent lines. Then $L_1 \cap L_2$ is a point.*

PROOF. The other possibility is that L_1 and L_2 coincide. This would lead to the equations

$$F(A, A) = 0, \qquad F(A, B) = 0, \qquad F(B, A) = 0, \qquad F(B, B) = 0.$$

In other words, the complex 2-planes

$$P_1 = \{Z : F(A, Z) = 0\}$$

and

$$P_2 = \{Z : F(B, Z) = 0\}$$

would be both equal to

$$\operatorname{span}_{\mathbb{C}}\{A, B\}.$$

But this would imply that A and B are dependent, i.e., $p = q$.

LEMMA 11. *Let p and q be distinct points of $Q \subset \mathbb{CP}^2$ and let A_0 and A_2 be vectors in $\mathbb{C}^3$ with $[A_0] = p$ and $[A_2] = q$. Then all solutions to*

$$F(A_0, B) = 0, \qquad F(A_2, B) = 0$$

project to the same point in Q and this point is the unique intersection of the complex tangent lines at p and q.

PROOF. Any solution to the equations projects to a point on both tangent lines and this point is unique.

It is possible to go further and find all the other points of Q whose tangent lines also go through the intersection point in this lemma. We digress to explain this. Let $[B_0]$ and $[B_1]$ be any two distinct points of $Q \subset \mathbb{CP}^2$ and let L be the complex line they determine. We say that a point $[B]$ is the polar point of L with respect to Q if

$$F(B_0, B) = 0, \qquad F(B_2, B) = 0.$$

Note that $[B]$ cannot lie on Q. The point $[B]$ depends only on L and not on the points B_0 and B_2.

LEMMA 12. *If $[A_0]$ and $[A_2]$ are also distinct points of $L \cap Q$ and if A satisfies*

$$F(A_0, A) = 0, \qquad F(A_2, A) = 0,$$

then $A = \mu B$ and so $[A] = [B]$.

PROOF. Let P be the complex 2-plane in $\mathbb{C}^3$ determined by L. In particular, P is determined as the plane through the points $0, B_0, B_2$. There is a vector E such that the equation of P is

$$\overline{E} \cdot C^{\mathrm{T}} Z = 0$$

with C as before. In particular, we have $\overline{E} \cdot C^T B_0 = 0$ and $\overline{E} \cdot C^T B_2 = 0$. Thus E is uniquely determined up to a scalar multiple by

$$(8) \qquad\qquad F(B_0, E) = 0, \qquad F(B_2, E) = 0,$$

and if we replace B_0 and B_2 by any two other vectors A_0 and A_2 with $[A_0]$ and $[A_2]$ distinct points of $Q \cap L$ we would obtain the same E (up to scalar multiple) since P would be unchanged. Since $E = \mu B$ we are done.

A complex line L either intersects Q in a curve or else is tangent to Q. We know what the polar is in the first case. In the second, we say that the polar of L is the point of tangency. So if $[B]$ is the polar of L then $[B] \in L$ if and only if L is tangent to Q at $[B]$. There is also the dual concept. We could start with a point $[E]$ and define the line L polar to $[E]$ with respect to Q as

$$L = \{[Z]: \overline{E} \cdot C^T Z = 0\}.$$

We have already seen that (8) defines the point $[E]$ which is the intersection of the complex lines tangent to Q at $[B_0]$ and $[B_2]$. Thus we have the following result.

LEMMA 13. (a) *Let L be a complex line in $\mathbb{CP}^2$. The complex lines tangent to Q at some point of $Q \cap L$ all intersect in a point which is the polar point of L with respect to Q.*

(b) *Let p be a point in $\mathbb{CP}^2$ but not on Q. The set of points $q \in Q$ which have the property that the complex line determined by q and p is tangent to Q is given by the intersection of Q with a line which is the polar line of p.*

REMARK. It is easy to see that $Q \cap L$ is a circle. For the map of Q to S^3 takes $Q \cap L$ to $S^3 \cap L'$ and restricts to a projective map of L to L'. This map preserves circles and since $S^3 \cap L'$ is clearly a circle, so is $Q \cap L$.

EXAMPLE. The polar of the line determined by $[1, 0, 0]$ and $[0, 0, 1]$ is $[0, 1, 0]$. In $\mathbb{C}^2$ this is the point at infinity which is on any complex line tangent to Q at some point $(0, u)$, u real.

Now let us return to the equations in Lemma 9. We have seen that

$$F(A_0, A_0) = 0 \quad \text{and} \quad F(A_2, A_2) = 0$$

mean that $[A_0]$ and $[A_2]$ are on Q, and that

$$F(A_0, A_1) = 0 \quad \text{and} \quad F(A_2, A_1) = 0$$

mean that $[A_1]$ is the intersection of the complex tangent lines at $[A_0]$ and $[A_1]$ and is the polar of the line determined by $[A_0]$ and $[A_1]$. The remaining equations are

$$(9) \qquad\qquad F(A_1, A_1) = 1 \quad \text{and} \quad F(A_2, A_0) = i,$$

and the determinant condition

$$(10) \qquad\qquad \det(A_0, A_1, A_2) = 1.$$

We have seen that once $[A_0]$ and $[A_2]$ are chosen, then $[A_1]$ is determined. But A_0, A_1, and A_2 are themselves only determined up to scalar multiples.

LEMMA 14. *Let* A_0, A_1, A_2 *be three independent vectors in* $\mathbb{C}^3$ *chosen so that each of the following quantities is zero*:

(11) $F(A_0, A_0), \qquad F(A_2, A_2), \qquad F(A_0, A_1), \qquad F(A_2, A_1).$

There exist unique nonzero complex numbers λ_1 *and* λ_2 *so that*

(12) $$F(\lambda_2 A_2, A_0) = i,$$

(13) $$F(\lambda_1 A_1, \lambda_1 A_1) = 1,$$

(14) $$\det(A_0, \lambda_1 A_1, \lambda_2 A_2) = 1.$$

PROOF. We have, assuming (11),

$$A^T C \overline{A} = \begin{pmatrix} 0 & 0 & F(A_0, A_2) \\ 0 & F(A_1, A_1) & 0 \\ F(A_2, A_0) & 0 & 0 \end{pmatrix},$$

and so

(15) $$|\det A|^2 = |F(A_0, A_2)|^2 F(A_1, A_1).$$

Since the vectors are independent, the left-hand side is nonzero. So from (12) and (14) we have

(16) $$\lambda_2 = i/F(A_2, A_0)$$

and

(17) $$\lambda_1 = -iF(A_2, A_0)/\det(A_0, A_1, A_2).$$

It is easily verified that (13) is now automatic.

Thanks to this lemma and our characterization of tangent lines we now have the result:

THEOREM 5. *Let* p *and* q *be distinct points of* $Q \subset \mathbb{CP}^2$ *and let* m *be the intersection of the complex tangents to* Q *at* p *and* q. *Let* A_0 *be some vector in* $\mathbb{C}^3$ *with* $[A_0] = p$. *There exist unique vectors* A_1 *and* A_2 *in* $\mathbb{C}^3$ *such that* $[A_2] = q$, $[A_1] = m$, *and* $A = (A_0, A_1, A_2)$ *is an element of* $\mathrm{SU}(2, 1)$. *Conversely, if* $B = (B_0, B_1, B_2)$ *is an element of* $\mathrm{SU}(2, 1)$ *then* $[B_0]$ *and* $[B_2]$ *are distinct points of* Q, *and* $[B_1]$ *is the point of intersection of the corresponding tangent lines.*

We have three degrees of freedom in choosing each of p and q and two in choosing A_0. Thus again we see that $\dim \mathrm{SU}(2, 1) = 8$. The action of $\mathrm{SU}(2, 1)$ on $\mathbb{C}^2$ is not effective since $\lambda(\mathrm{id})$ acts as the identity when $\lambda^3 = 1$. But the dimension of $\mathrm{SU}(2, 1)$ as a group acting on $\mathbb{C}^2$ is still eight. Also if we fix p, then five choices remain and we see again that the isotropy group has dimension five. Note that A maps $(1, 0, 0)$ to A_0 and so as a map of

Q, A maps 0 to $[A_0]$. In particular, $\mathrm{SU}(2, 1)$ is transitive on Q. This also follows from Lemma 6 since $\mathrm{U}(2, 1)$ and $\mathrm{SU}(2, 1)$ give the same set of biholomorphisms of $\mathbb{CP}^2$.

In Chapter 4, we will need the result that given two complex lines in $\mathbb{C}^2$, intersecting Q in curves, there is an automorphism of Q which takes one line to the other. We work in $\mathbb{CP}^2$.

COROLLARY. *Given any two complex projective lines L_1 and L_2 in $\mathbb{CP}^2$ such that $L_1 \cap Q$ and $L_2 \cap Q$ each contain more than one point, there is a four-parameter family of elements in $\mathrm{SU}(2, 1)$ which take L_1 to L_2.*

PROOF. Let L be the line determined by the points $[1, 0, 0]$ and $[0, 0, 1]$. It is enough to find those elements of $\mathrm{SU}(2, 1)$ which take L to L_1. So let p and q be two distinct points in $L_1 \cap Q$ and let A_0 be a vector in $\mathbb{C}^3$ with $[A_0] = p$. By the theorem there exist unique vectors A_1 and A_2 such that

$$A = (A_0, A_1, A_2) \in \mathrm{SU}(2, 1),$$

$$[A_2] = q.$$

Clearly A takes Q to itself and L to L_1. Note that $L \cap Q$ is a real curve, so the same is true for $L_1 \cap Q$. Now we count parameters. Both p and q are on this curve. This gives 2 parameters. In place of A_0 we could take any nonzero multiple. This gives 2 more parameters. Thus there are four parameters.

The following exercises are easy applications of the theorem. We will need the second one for our next corollary.

EXERCISE 3. Show that there are precisely three elements of $\mathrm{SU}(2, 1)$ which map $(1, 0, 0)$ and $(0, 0, 1)$ to multiples of themselves and map $(-1, \sqrt{2}, -i)$ to a multiple of $(1, 1 - i, i)$. Then find the unique automorphism of Q that leaves $[1, 0, 0]$ and $[0, 0, 1]$ fixed and maps $[-1, \sqrt{2}, -i]$ to $[1, 1 - i, i]$. (Hint: The first maps must have diagonal matrices.)

EXERCISE 4. Find a biholomorphism of S^3 to Q that takes

(18) $\{(z_1, z_2)\colon \operatorname{Im} z_1 = \operatorname{Im} z_2 = 0\} = \{x^2 + u^2 = 1, y = 0, v = 0\}$

to

(19) $z_1 = (1 - i)t, \qquad z_2 = it^2.$

(Hint: Multiply by i to take (18) to

(20) $y^2 + v^2 = 1, \qquad x = 0, \qquad u = 0,$

and then use the maps in Lemma 7, equation (2) and the above exercise.)

EXERCISE 5. Show that the group of automorphisms of Q that leave the origin fixed and map the u-axis to itself is given by

$$\begin{pmatrix} Re^{-i\theta/2} & 0 & re^{-i\theta/2} \\ 0 & e^{i\theta} & 0 \\ 0 & 0 & \dfrac{1}{R}e^{-i\theta/2} \end{pmatrix},$$

where R, r, and θ are real. (Hint: First convince yourself that four of the zeros are correct. Then choose the other components to obtain an element of $SU(2, 1)$.)

Here is a result which will be needed in Chapter 9. Indeed, we essentially reprove it there, but from a different point of view. For convenience in the statement of our result we consider $SU(2, 1)$ as acting on S^3. For convenience in the proof, we soon switch to its action on Q.

COROLLARY. *The subgroup of S^3 which maps the curve (18) to itself is of dimension three.*

PROOF. The linear map with matrix

(21) $$\begin{pmatrix} a & -b \\ b & a \end{pmatrix}, \qquad a, b \text{ real}, a^2 + b^2 = 1,$$

maps (18) to itself and takes $(1, 0)$ to (a, b). Thus we need to show that the subgroup of $SU(2, 1)$ which maps (18) to itself and which fixes some given point is of dimension two. We now switch to the action of $SU(2, 1)$ on Q. We have just seen, in Exercise 3, that there is a biholomorphism (except at one point) which takes S to Q and takes (18) to (19). We look for elements of $SU(2, 1)$ which map (19) to itself and fix the origin. Naturally, it is convenient to use homogeneous coordinates. The curve (19) is then given by

(22) $$\{[1 + i, 2t, -(1 - i)t^2]: t \in \mathbb{R}\}.$$

Let $P = [1, 0, 0]$, that is, $p = (0, 0)$. Let q be any other point on (19) obtained by fixing some value of t in (22); so $q = [1 + i, 2t, -(1 - i)t^2]$. The point m, as in the theorem, is given by $m = [1, \frac{1}{2}(1 - i)t, 0]$. Now let

$$A_0 = \lambda(1 + i, 0, 0),$$

$$A_1 = (1, \frac{1}{2}(1 - i)t, 0),$$

and

$$A_2 = (1 + i, 2t, -(1 - i)t^2),$$

where λ is to be chosen later.

By the theorem there are unique numbers λ_1 and λ_2 (depending of course on λ) so that

$$A = (A_0, \lambda_1 A_1, \lambda_2 A_2)$$

is in $SU(2, 1)$. We claim that there is a one-real-parameter family of choices for λ so that A maps the curve (19) to itself. The condition that A maps (22) to itself is that for every T there exist $\mu \in \mathbb{C}$ and $S \in \mathbb{R}$ such that

(23)
$$A \begin{pmatrix} 1+i \\ 2T \\ -(1-i)T^2 \end{pmatrix} = \mu \begin{pmatrix} 1+i \\ 2S \\ -(1-i)S^2 \end{pmatrix}.$$

Since
$$F(A_2, A_0) = -2t^2\bar{\lambda}$$

and
$$\det(A_0, A_1, A_2) = (1-i)\lambda t^3,$$

we obtain from (16) and (17)
$$\lambda_1 = \frac{\bar{\lambda}}{\lambda}\frac{1}{t}(i-1)$$

and
$$\lambda_2 = -\frac{i}{2t^2\bar{\lambda}}.$$

So (23) is equivalent to
$$\mu = \lambda(1+i) + (2i\bar{\lambda}/(\lambda t)) + ((1+i)T^2/(2t^2\bar{\lambda})),$$
$$\mu S = (i\bar{\lambda}T/\lambda) + (1+i)T^2/(2\bar{\lambda}t),$$
$$\mu S^2 = (1+i)T^2/(2\bar{\lambda}).$$

We may obtain $\mu^2 S^2$ by multiplying the first and third equations or by squaring the second equation. The results must be the same and equating them gives an identity in T. In fact, the coefficients on each side of T^3 and T^4 are identical. From the coefficients of T^2, we obtain
$$\lambda^3 = i\bar{\lambda}^3.$$

This fixes the argument of λ but leaves $|\lambda|$ arbitrary. Thus the subgroup of $SU(2, 1)$ leaving (18) (or (19)) fixed depends on three parameters—one specifies the image of $(0, 0)$, i.e., composition with some matrix (21) leaves $(0, 0)$ fixed; one specifies the image q of the point at infinity; and one fixes the modulus of λ.

We will end this chapter with an explicit representation for the elements of $SU(2, 1)$. First fix any particular element of $SU(2, 1)$, say $A = (A_0, A_1, A_2)$.

LEMMA 15. *Let* τ, λ, μ *be complex numbers,* τ *nonzero with* $\tau^3 = \lambda^2\bar{\lambda}$. *Let* p *be a real number. The matrix* $B = (B_0, B_1, B_2)$ *is an element of* $SU(2, 1)$ *when*

$$B_0 = \tau A_0,$$
$$B_1 = \frac{\tau}{\lambda}(A_1 + i\bar{\mu}A_0),$$
$$B_2 = \frac{\tau}{|\lambda|^2}(A_2 - \mu A_1 - (p + \frac{1}{2}i|\mu|^2)A_0),$$

and every element of $SU(2, 1)$ *that maps the origin to* $[A_0]$ *is of this form.*

PROOF. It is easy to verify that B satisfies our equations

$$F(B_1, B_1) = 1, \qquad F(B_2, B_0) = i, \qquad F(B_0, B_2) = -i,$$
$$F(B_i, B_k) = 0 \text{ otherwise},$$
$$\det(B_0, B_1, B_2) = 1,$$

and so does give an element of $SU(2, 1)$.

Next let $B = (B_0, B_1, B_2)$ be an element of $SU(2, 1)$ that maps the origin to $[A_0]$. So we must have $[B_0] = [A_0]$ and so $B_0 = \tau A_0$ for some τ. Further we know that $[B_1]$ lies on the complex tangent line through $[A_0]$ as does $[A_1]$. Thus B_1 is in the complex plane spanned by A_0 and A_1. Thus B_1 is a combination of A_0 and A. We choose to write this as

$$(24) \qquad\qquad B_1 = \frac{\tau}{\lambda}(A_1 + i\overline{\mu}A_0).$$

We also know that $[B_1]$ lies on the complex tangent line through $[B_2]$. We write this as

$$(25) \qquad\qquad F(B_2, B_1) = 0.$$

If we let

$$B_2 = \alpha A_0 + \beta A_1 + \gamma A_2,$$

and substitute this along with (24) into (25) we see that $\beta = -\mu\gamma$; if we substitute this expression for B_2 into

$$F(B_2, B_0) = i,$$

we obtain

$$\gamma = \frac{1}{\overline{\tau}}.$$

From

$$F(B_2, B_2) = 0$$

we obtain

$$\alpha = -\frac{1}{\overline{\tau}}\left(\rho + \frac{1}{2}i|\mu|^2\right)$$

where ρ is arbitrary. Finally

$$\det(B_0, B_1, B_2) = 1$$

leads to

$$\tau^3 = \lambda^2\overline{\lambda}.$$

So each B is of the claimed form.

If, for example, we take A to be the identity then we get for B,

$$(26) \qquad B = \begin{pmatrix} \tau & i\tau\overline{\mu}/\lambda & -(\rho + \frac{1}{2}i|\mu|^2)\tau/|\lambda|^2 \\ 0 & \tau/\lambda & -\mu\tau/|\lambda|^2 \\ 0 & 0 & \tau/|\lambda|^2 \end{pmatrix}.$$

This is the isotropy subgroup (of the origin) of $SU(2, 1)$ acting on $Q \subset \mathbb{CP}^2$.
It gives the set of all CR automorphisms of Q that leave the origin fixed.
The same set of automorphisms is given by the matrices (we are writing C
for $\frac{1}{\lambda}$ and changing ρ to s)

$$\begin{pmatrix} 1 & i\bar{\mu}C & -(s + \frac{1}{2}i|\mu|^2)|C|^2 \\ 0 & C & -\mu|C|^2 \\ 0 & 0 & |C|^2 \end{pmatrix}.$$

Now let $a = -\mu\bar{C}$, $r = -s|C|^2$, and $\rho = |C|^2$. The result is

(27)
$$\begin{pmatrix} 1 & -i\bar{a} & r - \frac{1}{2}i|a|^2 \\ 0 & C & aC \\ 0 & 0 & \rho \end{pmatrix}$$

with a and C arbitrary complex numbers but $C \neq 0$, r an arbitrary real
number and $\rho = |C|^2$. Any automorphism of $v = \frac{1}{2}|z|^2$ which leaves the
origin fixed is given by such a matrix. Compare this to (1) which gives the
automorphisms of $v = |z|^2$ leaving the origin fixed.

We end our discussion with an explicit representation of the finite elements
of $SU(2, 1)$, i.e., of those elements that leave infinity fixed. To do this, we
start with three vectors in $\mathbb{C}^3$ with

(28) $F(A_0, A_0)$, $F(A_0, A_1)$, $F(A_2, A_1)$, $F(A_2, A_1)$

all equal to zero. Let

$$Q = \{(z, w): w = u + \frac{i}{2}|z|^2\},$$

and now fix some ζ and s. We take $[A_0] = (\zeta, s + \frac{i}{2}|\zeta|^2)$ and, for example,

$$A_0 = (1, \zeta, s + \frac{i}{2}|\zeta|^2).$$

The complex tangent line to Q at $[A_0]$ is spanned by $(1, i\bar{\zeta})$. Let us take
$[A_1]$ to be the point at infinity on this line. So

$$A_1 = (0, 1, i\bar{\zeta}).$$

The choice that leads to the simplest expression is to take $[A_2]$ to be the
point at infinity on Q. That is,

$$A_2 = (0, 0, 1).$$

Note that $F(A_2, A_1) = 0$ so $[A_1]$ is indeed on the tangent line to Q at
$[A_2]$. Also note that

$$F(A_1, A_1) = 1, \qquad F(A_2, A_0) = i, \quad \text{and} \quad \det(A_0, A_1, A_2) = 1.$$

Thus

(29)
$$\begin{pmatrix} 1 & 0 & 0 \\ \zeta & 1 & 0 \\ s + \dfrac{i}{2}|\zeta|^2 & i\bar{\zeta} & 1 \end{pmatrix}$$

is an element of $SU(2, 1)$ and the general finite element of $SU(2, 1)$ is obtained by multiplying this matrix on the right by the matrix given in (26).

CHAPTER 3

Formal Theory of the Normal Form

As we have seen, two hypersurfaces are generally inequivalent. This suggests the problem of finding invariants that distinguish the biholomorphism classes of hypersurfaces. In this chapter and the next we present one solution of this problem as developed by J. Moser. Then, in Chapter 6, we present quite a different approach which is due to E. Cartan.

Of course we already know one invariant—for there to exist a local biholomorphism taking (M, p) to (M', p') one must have that (M, p) and (M', p') are both either degenerate or nondegenerate. The more interesting of these cases is naturally the nondegenerate case, and from now on we limit ourselves to this. Moser posed this problem of finding the invariants in a somewhat different form: Is it possible to find a local biholomorphism ψ which takes (M, p) to $(N, 0)$ where N is as "nice" as possible. If one could find enough conditions on N so that the choice of ψ is unique then $(N, 0)$ would be called the normal form of (M, p) and we would have solved the "equivalence problem": There is a local biholomorphism taking (M, p) to (M', p') if and only if both (M, p) and (M', p') have the same normal form $(N, 0)$. In more concrete terms, the coefficients of the Taylor series of the function defining N as a graph over some fixed hyperplane would be the biholomorphism invariants. Unfortunately, conditions imposed on N cannot always determine ψ (or N) uniquely. But what Moser did accomplish is a reduction to an easier, indeed finite-dimensional, equivalence problem: (M, p) and (M', p') are locally biholomorphic if and only if $(N, 0)$ and $(N', 0)$ are locally biholomorphic, when $(N, 0)$ is some "nice" form for (M, p) and $(N', 0)$ is some "nice" form for (M', p'). In the next chapter we explain our claim that this constitutes a reduction to a finite-dimensional problem.

This is how we shall proceed. We start with a hypersurface M that is nondegenerate at some point p and consider all maps ϕ such that $\phi(M)$ is given as a graph over the (z, u) plane. Conditions are imposed on the coefficients given by the function associated to this graph with the aim of making the map ϕ unique. But this aim is not usually obtainable. For consider the hyperquadric Q given by $v = |z|^2$. The coefficients here are already as nice as possible. Thus if, in general, the map ϕ could be made unique, in this particular case it would have to be the identity. But of course any element of G_0 (the isotropy group at the origin of Q) achieves the same coefficients. So instead of making $\phi(M)$ unique we try to make it as "simple" as possible.

We already have a start towards developing a normal form since we know (Lemma 1.4) that if (M, p) is nondegenerate then it is equivalent to a hypersurface of the form

$$(1) \qquad v = a|z|^2 + Az^2 + \overline{A}\,\overline{z}^2 + bu^2 $$
$$ + Bzu + \overline{B}\overline{z}u + \cdots $$

with $a \neq 0$. We now try to simplify this expansion by finding biholomorphisms that make certain coefficients equal to zero. First we show that a can be taken to be one and A can be taken to be zero. To do this we implicitly define a biholomorphism $\phi(z, w) = (z^*, w^*)$ by

$$ z = z^* \quad \text{and} \quad w = aw^* + 2iAz^{*2}. $$

We have

$$(2) \qquad v = \operatorname{Im} w = \operatorname{Im}(aw^* + 2iAz^{*2}), $$

and so

$$(3) \qquad v = av^* + Az^{*2} + \overline{A}\overline{z}^{*2}. $$

Note that

$$ u = au^* + 2\operatorname{Re}(iAz^{*2}), $$

and so third-order terms in $|z| + |u|$ are also third-order in $|z^*| + |u^*|$. Thus (1) can be rewritten as

$$ v = a|z^*|^2 + Az^{*2} + \overline{A}\overline{z}^{*2} + ba^2u^{*2} $$
$$(4) \qquad + aBz^*u^* + a\overline{B}\overline{z}^*u^* + \cdots $$

with the dots denoting third-order and higher terms.

Comparing (3) and (4), we see that

$$(5) \qquad v^* = |z^*|^2 + abu^{*2} + 2\operatorname{Re}(Bz^*u^*) + \cdots . $$

EXERCISE 1. Adapt the transformation $(z, w) \to (z^*, w^*)$ in order to show that any hypersurface given by

$$ v = \operatorname{Re}(h(z) + zg(z, \overline{z}, u)) $$

can be transformed to one of the form

$$ v^* = \operatorname{Re}(z^* g^*(z^*, \overline{z}^*, u^*)) $$

whenever $h(z)$ is holomorphic. In particular, show that $v = \operatorname{Re}(h(z))$ is equivalent to the hyperplane $v^* = 0$.

EXERCISE 2. Consider a map

$$z = \alpha_1 z^* + \alpha_2 w^* + \cdots ,$$

$$w = \beta_1 z^* + \beta_2 w^* + \gamma_1 z^{*2} + \gamma_2 z^* w^* + \gamma_3 w^{*2} + \cdots ,$$

which takes one particular hypersurface of the form (5),

$$v = |z|^2 + 2\operatorname{Re}(Az)u + bu^2 + \cdots ,$$

to another of the same form,

$$v^* = |z^*|^2 + 2\operatorname{Re}(A^* z^*)u^* + b^* u^{*2} + \cdots .$$

Show that α_1, α_2, and γ_3 are arbitrary, β_1 and γ_1 are zero, $\gamma_2 = 2i\alpha_1\bar{\alpha}_2 + 2iA\alpha_1\beta_2$ and $\beta_2 = |\alpha_1|^2$. (Note that the results of this exercise are consistent with what we already know about the elements of G_0. See Chapter 2.)

It will be very useful to assign different "weights" to the variables z and u (and similarly to z and w).

DEFINITION. A monomial $z^a \bar{z}^b u^c$ is of *weight* $\nu = a+b+2c$. A monomial $z^a w^b$ is of *weight* $\nu = a + 2b$. In either case a polynomial of weight ν is a linear combination of monomials each of weight ν.

Thus a manifold of the form (5) can be written

(6) $$v = |z|^2 + \sum_{\nu \geq 3} F_\nu(z, \bar{z}, u)$$

where F_ν is a polynomial of weight ν.

The previous exercise shows, in particular, that if ϕ maps one hypersurface of the form (6) into another of the same form and preserves the origin then (reversing now the roles of z^* and z, etc.)

(7) $$z^* = Cz + \sum_{\nu \geq 2} \phi_\nu^{(1)}(z, w),$$

(8) $$w^* = \rho z + \sum_{\nu \geq 3} \phi_\nu^{(2)}(z, w),$$

with $\rho = |C|^2$.

EXERCISE 3. Show that, conversely, any biholomorphism given by (7) and (8) maps every hypersurface of the form (6) to another of the same form. Note that this implies that if Φ maps the hyperquadric $v = |z|^2$ to itself and preserves the origin then Φ maps every hypersurface of the form (6) to another of the same form. Finally, show directly that the biholomorphisms given by (7) and (8) form a group under composition. (This is meant formally—do not worry about the domains of convergence.)

We will now see that it is possible to define a normal form for nondegenerate hypersurfaces and that the map of such a surface into its normal form is unique up to (a complicated action of) the isotropy group G_0. At first we work only with formal power series. Fortunately, the convergence proof can be avoided by a careful examination of the geometric consequences of the normalizations. We will be using, in addition to the weight decomposition, a decomposition by "type".

DEFINITION. A term of the form $g(u)z^j \bar{z}^k$, where $g(u)$ is a formal power series, is of *type* (j, k).

We expand the defining function of our hypersurface:

$$v = F(z, \bar{z}, u) = \sum_{(j,k)} F_{jk}(u)z^j \bar{z}^k.$$

In choosing to work with partial Taylor expansions $F = \sum F_{j,k}(u)z^j \bar{z}^k$ instead of full Taylor expansions $F = \sum F_{j,k,m} z^j \bar{z}^k u^m$ one is guided by two facts: the hyperquadric is much easier to work with when it is realized as $v = |z|^2$ rather than as the unit sphere; and there are basic differences between the u and z variables which show up in various analytic results. But in the end, the partial expansion is justified because it reveals interesting properties which also arise when CR hypersurfaces are studied by a completely different method (cf. Chapters 6 and 7).

So let us start with a hypersurface M given by

$$(9) \qquad v = |z|^2 + \sum_{\nu \geq 3} F_\nu = |z|^2 + \sum F_{jk}(u)z^j \bar{z}^k,$$

and seek a biholomorphism ψ such that for $M^* = \psi(M)$ certain of the coefficients $F_{jk}^*(u)$ are identically zero. We shall also require that ψ be normalized and in this way we do obtain a unique ψ and a unique normal form. But nonuniqueness comes from the fact that there are many maps ϕ that put a given hypersurface into the form (9). Then for each such ϕ we obtain a unique ψ for which $\psi \circ \phi$ provides a map into normal form. Fortunately, as we show in the next chapter, the totality of maps $\psi \circ \phi$ is actually finite-dimensional and indeed bijective to G_0.

Note that the identity map in G_0 is uniquely specified if we consider $\psi = (\psi_1, \psi_2) \in G_0$ and require at the origin

$$(10) \qquad d\psi = I, \qquad \mathrm{Re}\frac{\partial^2 \psi_2}{\partial w^2} = 0 \quad (I \text{ is the } 2 \times 2 \text{ identity}).$$

Thus when we try to find a normal form for a hypersurface (9) it is natural to restrict the biholomorphisms by considering only those that satisfy (10).

THEOREM 1 (THEOREM 2.2 OF [CM]). *A formal hypersurface*

$$v = |z|^2 + \sum_{\nu \geq 3} F_\nu$$

can be transformed by a formal transformation

$$z^* = z + f(z, w), \qquad w^* = w + g(z, w)$$

into a formal hypersurface

$$v = |z|^2 + N_{24}(u)z^2\bar{z}^4 + N_{42}(u)z^4\bar{z}^2$$
$$+ \sum_{\substack{j+k\geq 7 \\ k\geq 2 \\ j\geq 2}} N_{jk}(u)z^j\bar{z}^k$$

(*stars have been suppressed*). *Further, this transformation* (*and hence also the coefficients* N_{jk}) *is made unique by the normalizations that at the origin* f, g, f_z, f_w, g_z, g_w, g_{zz}, *and* $\mathrm{Re}\, g_{ww}$ *are all zero* (*in the sense that each formal Taylor expansion has no constant term*).

REMARKS. 1) The condition $g_{zz}(0) = 0$ could be omitted since it is automatic (see Exercise 2); however, it is convenient to group it with the other normalizations.

2) In terms of weight decompositions the normalizations of the theorem become

(11)
$$f = \sum_{\nu\geq 2} f_\nu(z, w) \quad \text{with } f_2(0, w) = 0,$$

$$g = \sum_{\nu\geq 3} g_\nu(z, w) \quad \text{with } \mathrm{Re}\, g_4(0, w) = 0.$$

COROLLARY. *If a formal transformation*

$$z^* = F(z, w), \qquad w^* = G(z, w),$$
$$\text{with } F(0) = G(0) = 0, \qquad F_z(0) \neq 0,$$

takes the hyperquadric $v = |z|^2$ *to a formal hypersurface*

$$v = |z|^2 + N_{24}(u)z^2\bar{z}^4 + N_{42}(u)z^4\bar{z}^2$$
$$+ \sum_{\substack{j+k\geq 7 \\ k\geq 2 \\ j\geq 2}} N_{jk}(u)z^j\bar{z}^k,$$

then this formal transformation is actually an element of the isotropy group G_0 *and each* N_{jk} *is zero.*

REMARK. It follows that any local biholomorphism that takes one open subset of Q to another extends to a global automorphism of Q (perhaps taking some finite point to the point at infinity).

PROOF OF COROLLARY. Set $C = F_z(0)$. From Exercise 2 we know that $G_w = |C|^2$, $G_z = 0$, and $G_{zz} = 0$. (All derivatives are evaluated at the origin.) Define a and r by $F_w = Ca$, $\mathrm{Re}\, G_{ww} = -2|C|^2 r$, and let $\gamma \in G_0$ correspond to (C, a, r). Let $\psi = (F, G)$ be the given transformation and $\tilde{\psi} = \psi \circ \gamma^{-1}$. Of course $\tilde{\psi}$ maps the hyperquadric to the same normal form as ψ does. A simple computation shows that $\tilde{\psi}$ satisfies the normalizations of the theorem. Thus both $\tilde{\psi}$ and $\{N_{jk}\}$ are unique. Hence $\tilde{\psi}$ must be the identity and each N_{jk} must be zero. Thus $\psi = \gamma$ and we are done.

DEFINITIONS. 1. The set of *admissible formal power series* is

$$\mathscr{F} = \{F(z, \bar{z}, u): \text{ with no term of weight less than 3}\}.$$

2. The set of *admissible formal hypersurfaces* is

$$\mathscr{H} = \{v = |z|^2 + F: F \in \mathscr{F}\}.$$

3. The set of *formal normalizations* is

$$\mathscr{N} = \{N = \sum N_{jk}(u)z^j \bar{z}^k: N_{jk} \text{ is a formal power series in } u$$

and N_{jk} is the zero series when $j \in \{0, 1\}$ or $k \in \{0, 1\}$

or $(j, k) \in \{(2, 2), (2, 3), (3, 2), (3, 3)\}\}.$

4. The set of *formal normal forms* is

$$\mathscr{M} = \{v = |z|^2 + N: N \in \mathscr{N}\}.$$

5. The set of *admissible formal transformation functions* is

$$\mathscr{D} = \{h = (f, g): f \text{ and } g \text{ are formal power series which satisfy (11)}\}.$$

Using these definitions, Theorem 1 may be restated: Each element of $\mathscr{H}$ may be transformed into an element of $\mathscr{M}$ by one and only one formal transformation $z^* = z + f(z, w)$, $w^* = w + g(z, w)$ with (f, g) an element of $\mathscr{D}$.

The proof of Theorem 1 is based on a study of the operator $L: \mathscr{D} \to \mathscr{F}$ defined by

$$L(f, g) = \text{Re}\{2\bar{z}f(z, u + i|z|^2) + ig(z, u + i|z|^2)\}.$$

Note that $\mathscr{F}$ has a natural gradation by weight

$$\mathscr{F} = \bigcup_\nu \mathscr{F}_\nu, \qquad \mathscr{F}_\nu = \{F_\nu\},$$

and if we take as a gradation on $\mathscr{D}$

$$\mathscr{D} = \bigcup_\nu \mathscr{D}_\nu, \qquad \mathscr{D}_\nu = \{h_\nu = (f_{\nu-1}, g_\nu)\},$$

then L respects this gradation: $L(\mathscr{D}_\nu) \subset \mathscr{F}_\nu$.

DEFINITION. A linear subspace $\mathscr{N}'$ of $\mathscr{F}$ is called a *complement* to $L(\mathscr{D})$ if

1. $L(\mathscr{D}) \cup \mathscr{N}'$ spans $\mathscr{F}$, $L(\mathscr{D}) \cap \mathscr{N}' = \{0\}$ and
2. $\mathscr{N}' = \bigoplus \mathscr{N}'_\nu$, $\mathscr{N}'_\nu = \mathscr{N}' \cap \mathscr{F}_\nu$.

Note that the lowest weight in $\mathscr{F}$ is $\nu = 3$ and so this is also the lowest possible weight in $\mathscr{N}'$. In fact, as we shall see later, $L(\mathscr{D})$ includes all terms of weight three and so $\mathscr{N}'_3$ is empty.

We shall first show that any complement corresponds to a normal form and then show that the set $\mathscr{N}$ previously defined is one such complement.

LEMMA 1. *Let $\mathcal{N}'$ be a complement to $L(\mathcal{D})$. Any hypersurface in $\mathcal{H}$ can be transformed by one and only one transformation*

$$\phi = (z + f, w + g) \quad \text{with} \quad (f, g) \in \mathcal{D}$$

to a formal hypersurface of the form

$$v = |z|^2 + N, \qquad N \in \mathcal{N}'.$$

PROOF. We start with $v = |z|^2 + F(z, \bar{z}, u)$ and ask when it can be transformed into

$$(12) \qquad v^* = |z^*|^2 + F^*(z^*, \bar{z}^*, u^*)$$

by a transformation

$$(13) \qquad z^* = z + f(z, w), \qquad w^* = w + g(z, w).$$

Starting with $v^* = \operatorname{Im} w^*$ we see that

$$v^* = \operatorname{Im}(w + g(z, w)) = v + \operatorname{Im} g(z, u + iv)$$
$$= |z|^2 + F(z, \bar{z}, u) + \operatorname{Im} g(z, u + iv).$$

Thus (12) can be achieved precisely when there exists $(f, g) \in \mathcal{D}$ that satisfies

$$(14) \quad |z^*|^2 + F^*(z^*, \bar{z}^*, u^*) = |z|^2 + F(z, \bar{z}, u) + \operatorname{Im} g(z, u + i(|z|^2 + F))$$

after the substitution

$$(15) \quad z^* = z + f(z, u + i(|z|^2 + F)), \qquad u^* = u + \operatorname{Re} g(z, u + i(|z|^2 + F)).$$

Let us do this substitution; we obtain

$$(16) \qquad \operatorname{Re}\{2\bar{z}f + ig\} = F - F^* - |f|^2$$

where $f = f(z, u + i(|z|^2 + F))$, $g = g(z, u + i(|z|^2 + F))$, $F = F(z, \bar{z}, u)$, and $F^* = F^*(z + f, \bar{z} + \bar{f}, u + \operatorname{Re} g)$.

In the weight expansion of (16) the smallest weight that occurs is three. Let us show that we can choose f and g such that $F_3^* \in \mathcal{N}'$. We will then use induction to show we can choose f and g so as to achieve $F^* \in \mathcal{N}'$. For the terms of weight three in (16) we obtain

$$\operatorname{Re}(2\bar{z}f_2(z, u + i|z|^2) + ig_3(z, u + i|z|^2)) = F_3(z, \bar{z}, u) - F_3^*(z, \bar{z}, u).$$

Thus we want to solve

$$(17) \qquad Lh = F_3(z, \bar{z}, u) - F_3^*(z, \bar{z}, u).$$

There is a unique $F_3^* \in \mathcal{N}'$ for which this is solvable; namely, F_3^* must be the projection of F_3 into $\mathcal{N}'$ defined by the decomposition $\mathcal{F} = L(\mathcal{D}) \oplus \mathcal{N}'$. Here we make use of the fact that $\mathcal{N}'$ is graded (Condition (2) of the definition of a complement). Further, for this F_3^* there is only one h that solves (17).

LEMMA 2. *If $h \in \mathcal{D}$ and $Lh = 0$, then $h = 0$.*

The proof is given at the end of this chapter.

We now use induction to prove the lemma. Clearly it is enough to show that if we have

$$v = |z|^2 + \sum_{\nu \geq 3} F_\nu \quad \text{with} \quad F_\nu \in \mathcal{N}' \quad \text{for} \quad \nu = 3, \ldots, K,$$

then there is a map of the form

(18) $$z^* = z + f_K(z, w), \qquad w^* = w + g_{K+1}(z, w)$$

that achieves $v^* = |z^*|^2 + \sum_{\nu \geq 3} F_\nu^*$ with $F_{K+1}^* \in \mathcal{N}'$. This is so because for the transformation (18) one has $F_\nu^* = F_\nu$ for $\nu \leq K$. When the transformation (18) is used in (16) the terms of weight $K + 1$ give the equation

$$Lh = F_{K+1}(z, \bar{z}, u) - F_{K+1}^*(z, \bar{z}, u)$$

for $h = (f_K, g_{K+1}) \in \mathcal{D}$. As before, there is a unique F_{K+1}^* for which this equation has a solution and, as before, this solution is itself unique. This concludes the proof of Lemma 1. Thus any complement to $L(\mathcal{D})$ can be used to define a normal form.

We now want to show that for a complement of $L(\mathcal{D})$ we may take

$$\mathcal{N} = \Big\{ N = \sum N_{jk}(u) z^j \bar{z}^k : N_{jk} = 0$$
$$\text{for } j = 0, \ j = 1, \ k = 0, \ k = 1,$$
$$\text{or } (j, k) \in \{(2, 2), (2, 3), (3, 2), (3, 3)\} \Big\}.$$

Clearly $\mathcal{N} = \bigoplus(\mathcal{N} \cap \mathcal{F}_\nu)$ so we concern ourselves only with (1) of the definition of complement. Let

(19) $\quad S = \{(j, 0), (j, 1), (0, k), (1, k), (2, 2), (2, 3), (3, 2), (3, 3)\}$

where j and k take on all nonnegative integer values. For any formal power series $F(z, \bar{z}, u) = \sum F_{jk}(u) z^j \bar{z}^k$ let $\pi F = \sum_{(j,k) \in S} F_{jk}(u) z^j \bar{z}^k$. To show that $\mathcal{F} = L(\mathcal{D}) \oplus \mathcal{N}$ we only need show that for any $F \in \mathcal{F}$ there is some $h \in \mathcal{D}$ with $\pi Lh = \pi F$. And to show that $L(\mathcal{D}) \cap \mathcal{N} = \{0\}$ we only need to show that $Lh = 0$ if $\pi Lh = 0$. In fact we shall see that $\pi Lh = 0$ even implies $h = 0$. This is a strengthened form of Lemma 2.

So given $F \in \mathcal{F}$ we seek $h \in \mathcal{D}$ for which $(Lh)_{(j,k)} = F_{(j,k)}$ for each $(j, k) \in S$. Further we want to show that such h is unique. Recall that $h = (f, g)$ where f and g satisfy the normalizations of (11). We no longer have need of weight expansions, so instead we write

(20) $\qquad f(z, w) = \sum f_k(w) z^k, \qquad g(z, w) = \sum g_k(w) z^k,$

and for the normalizations we have that

(21) $\qquad\qquad f_0, \ f_0', \ f_1, \ g_0, \ g_0', \ \mathrm{Re}\, g_0'', \ g_1, \ g_2$

are all zero at $w = 0$.

Here and below the primes denote derivatives with respect to w. Thus for any function $\phi(z, w)$ we have the expansion

$$\phi(z, u + i|z|^2) = \phi(z, u) + i\phi'(z, u)|z|^2 - \frac{1}{2}\phi''(z, u)|z|^4$$
$$- \frac{i}{6}\phi'''(z, u)|z|^6 + \cdots .$$

In particular, for a term $\phi_k(w)z^k$ we have

$$\phi_k(u + i|z|^2)z^k = \phi_k(u)z^k + i\phi'_k(u)z^{k+1}\overline{z} - \frac{1}{2}\phi''_k(u)z^{k+2}\overline{z}^2$$
$$- \frac{i}{6}\phi'''_k(u)z^{k+3}\overline{z}^3 + \cdots .$$

We now expand $2Lh$:

$$2Lh = 2\,\mathrm{Re}\{2\overline{z}f(z, u + i|z|^2) + ig(z, u + i|z|^2)\}$$
$$= 2\overline{z}f + 2z\overline{f} + ig - i\overline{g}$$
$$= \sum_k (2f_k z^k \overline{z} + 2if'_k z^{k+1}\overline{z}^2 - f''_k z^{k+2}\overline{z}^3 + \cdots)$$
$$+ \sum_k (2\overline{f}_k \overline{z}^k z - 2i\overline{f'_k}\overline{z}^{k+1}z^2 - \overline{f''_k}\overline{z}^{k+2}z^3 + \cdots)$$
$$+ \sum_k (ig_k z^k - g'_k z^{k+1}\overline{z} - \frac{i}{2}g''_k z^{k+2}\overline{z}^2$$
$$+ \frac{1}{6}g'''_k z^{k+3}\overline{z}^3 + \cdots)$$
$$+ \sum_k (-i\overline{g}_k \overline{z}^k - \overline{g'_k}\overline{z}^{k+1}z + \frac{i}{2}\overline{g''_k}\overline{z}^{k+2}z^2$$
$$+ \frac{1}{6}\overline{g'''_k}\overline{z}^{k+3}z^3 + \cdots).$$

(The argument of each f_k, g'_k, etc. is u.)

We list the terms in $2Lh$ of type (k, j) for each $(k, j) \in S$. Note that we need only consider $k \geq j$ since $\mathscr{F}$ consists of real power series.

$(0, 0): -2\,\mathrm{Im}\, g_0$,

$(1, 0): 2\overline{f}_0 + ig_1$,

$(k, 0)$ for $k \geq 2$: ig_k,

$(1, 1): 4\,\mathrm{Re}\, f_1 - 2\,\mathrm{Re}\, g'_0$,

$(2, 1): 2f_2 - 2i\overline{f'_0} - g'_1$,

$(k, 1)$ for $k \geq 3$: $2f_k - g'_{k-1}$,

$(2, 2): -4\,\mathrm{Im}\, f'_1 + \mathrm{Im}\, g''_0$,

$(3, 2): 2if'_2 - \overline{f''_0} - \frac{i}{2}g''_1$,

$(3, 3): -2\,\mathrm{Re}\, f''_1 + \frac{1}{3}\,\mathrm{Re}\, g'''_0$.

We need to show that for each $F \in \mathscr{F}$ there is a unique $h = (f, g) \in \mathscr{H}$ that satisfies $(2Lh)_{(k,j)} = F_{(k,j)}$ for $(k, j) \in S$. First we note that

$$g_k(u) = -2iF_{(k,0)}(u) \quad \text{for } k \geq 2,$$

and

$$f_k(u) = F_{(k,0)}(u) + g'_{k-1}(u) \quad \text{for } k \geq 3.$$

The normalization $g_2(0) = 0$ is automatic since F begins with weight three.

Now look at those equations that contain $\overline{f}_0$:

$$2\overline{f}_0 + ig_1 = 2F_{(1,0)},$$

$$2\overline{f}_0' + 2if_2 - ig_1' = 2iF_{(2,1)},$$

$$\overline{f}_0'' - 2if_2' + \frac{i}{2}g_1'' = -2F_{(3,2)}.$$

Differentiate the first equation twice and the second once and then eliminate f_2' and g_1''. The result is

$$4\overline{f}_0'' = 2iF_{21}' - 2F_{32} + F_{10}''.$$

This ordinary differential equation, with the initial conditions given by the normalization (21), completely determines f_0. Now return to the first equation and then the second equation in order to determine $g_1(u)$ and $f_2(u)$.

The $(0,0)$ equation determines $\operatorname{Im} g_0$ and then the $(2,2)$ equation and the normalization (21) determine $\operatorname{Im} f_1$. From the $(1,1)$ equation and the $(3,3)$ equation we may derive

$$\operatorname{Re} g_0''' = -\frac{3}{2}(F_{(1,1)}'' + 2F_{(3,3)}),$$

and again (21) provides the initial conditions needed to determine $\operatorname{Re} g_0(u)$ uniquely. Finally, $(1,1)$ now allows us to determine $\operatorname{Re} f_1$ and we are done. In particular, if F is zero then h must also be zero. This is the promised proof of Lemma 2.

EXERCISE 4. Actually we are not done. The above computations show that if there is an h with $Lh_{(k,j)} = F_{(k,j)}$ for $(k, j) \in S$ then h is uniquely determined. One should now work backwards and show that the h so determined actually satisfies all the equations. Or one could try a more general viewpoint. Note that if we ignore the equations $(k, 0)$ for $k \geq 2$ and $(k, 1)$ for $k \geq 3$ and if we take as unknowns f_0, f_1, f_2, g_0, g_1, then we have ten real unknowns and ten real equations, some of which are algebraic and some of which are differential. We also have some initial conditions which guarantee that if a solution exists then it must be unique. Presumably there is some general technique that allows us to verify that these initial conditions are also minimal in the sense that the solution does in fact exist.

CHAPTER 4

Geometric Theory of the Normal Form

In this chapter we go through in detail the lowest-dimensional case of §3 of [**CM**]. Our aim is to show that the normalizations in Chapter 3 used to determine the formal transformations into normal form correspond to a geometric structure on the hypersurface. A consequence is that the formal transformations actually converge. Thus we shall simultaneously find a rich geometric structure and avoid a convergence proof for formal power series.

For typographical convenience, in this chapter we use $H^{1,0}$ for vectors, rather than one-forms, and set

$$H^{1,0} = \overline{V} = \{X - iJX : X \in H\}.$$

Recall that H denotes the characteristic two-plane distribution. We say that a curve $\gamma(t)$ is transverse to H if the tangent vector $\gamma'(t)$ is never in H. The following six propositions show how to determine a unique map into normal form starting with the choices of some data at one point. We work as usual with a nondegenerate hypersurface M and some point p on M and understand that all biholomorphisms are local.

PROPOSITION 1. *Given a real analytic parametrized curve* $\gamma(t)$ *in* M *with* $\gamma(0) = p$ *and* γ *transverse to* H, *there are biholomorphisms taking* M *to the form*

$$v = F_{11}(u)|z|^2 + \sum_{\substack{k \geq 2 \\ j \geq 2}} F_{kj}(u)z^k \bar{z}^j$$

and $\gamma(t)$ *to* $(0, t)$.

PROPOSITION 2. *Let* $e(t)$ *be any real analytic section of* $H^{1,0}$ *along* γ. *There is a unique biholomorphism from Proposition 1 that takes* $e(t)$ *to* $\frac{\partial}{\partial z}\big|_{(0,t)}$.

PROPOSITION 3. *There is a unique positive function* $\rho(t)$ *such that if the vector field* $e(t)$ *satisfies* $|e(t)| = \rho(t)$, *then* $F_{11}(u) = \pm 1$.

PROPOSITION 4. *Given a unit vector* $v_0 \in H_p^{1,0}$ *there is a unique unit section* $v(t)$ *of* $H^{1,0}$ *along* γ *such that if* $e(t) = \rho(t)v(t)$, *then* $F_{22}(u) = 0$. *The* section $\nu(t)$ is determined by solving a first-order ordinary differential equation with initial data $\nu(0) = \nu_0$.

PROPOSITION 5. *There is a projective family of parametrizations for γ such that if t is one of these parametrizations, then, in addition to the above, one has that $F_{33}(u) = 0$. The parametrization is obtained by solving a third-order ordinary differential equation.*

PROPOSITION 6. *For each direction transverse to H at p there is a unique (unparametrized) real analytic curve through p and tangent to that direction such that there exists some biholomorphism taking M to*

$$v = |z|^2 + \sum_{\substack{k \geq 2 \\ j \geq 2}} F_{kj}(u) z^k \bar{z}^j, \quad \text{with} \quad F_{32}(u) = 0,$$

and γ to the u-axis. This curve is determined by solving a second-order ordinary differential equation. Further, for any real analytic parametrization and any $e(0)$, with $|e(0)| = \rho(0)$, the unique biholomorphism given by Propositions 1 to 4 also satisfies $F_{32}(u) = 0$. In particular, if one of the parametrizations of Proposition 5 is chosen we have a unique map to normal form.

THEOREM 1. *Let p be a point on a nondegenerate hypersurface M, and Γ a direction at p transverse to H_p, and e_0 a vector in $H_p^{1,0}$. Let γ be the unique curve from Proposition 6 in the direction Γ and let $\gamma(t)$ be a parametrization of γ by one of the parametrizations from Proposition 5. There exists a unique local biholomorphism Φ defined in a neighborhood of p satisfying*

(i) $\Phi(M)$ *is in normal form*

$$v = |z|^2 + \sum_{\substack{k \geq 2 \\ j \geq 2}} F_{kj}(u) z^k \bar{z}^j, \quad F_{22}(u) = F_{32}(u) = F_{33}(u) = 0,$$

(ii) $\Phi(\gamma(t)) = (0, t)$,

(iii) $\varphi_* e_0$ *is a positive multiple of $\frac{\partial}{\partial z}\big|_0$.*

Further, if ψ is any local biholomorphism such that $\psi(p) = 0$ and $\psi(M)$ is in normal form then $\psi^{-1}(0, t)$ is one of the curves in Proposition 6 parametrized according to Proposition 5.

REMARKS.

1. Proposition 4 implies that we have parallel translation of vectors in H along any (real analytic) curve transverse to H.

2. Proposition 5 means that there is a map $t: \mathbb{R}^1 \to \gamma$, $\gamma(0) = p$, such that the unique biholomorphism that takes $\gamma(t)$ to $(0, t)$ and that satisfies Propositions 2, 3, and 4 also satisfies $F_{33}(u) = 0$. Further, a second parametrization $s: \mathbb{R}^1 \to \gamma$ has this same property if and only if $s = \frac{t}{at+b}$. Such projective parametrizations are discussed in detail in the next chapter.

3. With Proposition 6 in mind, we include in the proofs of Propositions 2 and 5 the fact that if $F_{32}(u) = 0$ then also $F_{32}^*(u^*) = 0$ for the image under the obtained biholomorphism.

PROOF OF PROPOSITION 1. We may take $p = 0$ and $H_0 = \{x, y\}$-plane. The given curve is $\gamma(t) = (a(t), b(t))$ for complex-valued, real analytic functions a and b. We have $\gamma(0) = (0, 0)$. That γ is transverse to H means that $b'(0) \neq 0$. Consider the transformation

$$z = z^* + a(w^*),$$
$$w = b(w^*).$$

This transformation is nonsingular at the origin, takes the u-axis to the curve γ, with $(0, u)$ going to $\gamma(u)$, and leaves H_0 unchanged. So the inverse transformation brings M to the form (written with * deleted)

(1) $v = F(z, \bar{z}, u)$ with $F(0, 0, u) = 0$,   and $F_z(0, 0, 0) = 0$.

EXERCISE 1. Show that manifolds of this form are preserved by all maps of the type

$$z^* = \alpha z + f(z, w),$$
$$w^* = w + g(z, w),$$

where α is a nonzero constant, f and g start with second-order terms, $f(0, w) = 0$, and $g(0, w) = 0$.

Basically, what we shall do is to eliminate all z^k terms in the expansion of (1) by use of an appropriate biholomorphism and then, by means of another biholomorphism, absorb all $z^k \bar{z}$ terms into F_{11}.

So we start with a transformation of the form

$$z^* = w,$$
$$w^* = w + g(z, w).$$

The $g(z, w)$ that we find will satisfy $g(0, w) = 0$ and so by the exercise will preserve the form (1).

Now

$$v^* = v + \frac{1}{2i}(g(z, u + iF(z, \bar{z}, u)) - \bar{g}(\bar{z}, u - iF(z, \bar{z}, u)))$$
$$= F^*(z^*, \bar{z}^*, u^*).$$

Let us temporarily consider z, $\bar{z}$, u as three independent complex variables. Then we seek a transformation that results in $F^*(z, 0, u^*) = 0$. That is,

$$0 = F(z, 0, u) + \frac{1}{2i}(g(z, u + iF(z, 0, u)) - \bar{g}(0, u - iF(z, 0, u))).$$

We may assume that $\bar{g}(0, w) = 0$, as long as the $g(z, w)$ we finally obtain satisfies $g(0, w) = 0$. So we seek $g(z, w)$ such that

$$g(z, u + iF(z, 0, u)) = -2iF(z, 0, u).$$

Let $s = u + iF(z, 0, u)$. Then $u = h(z, s)$ with $h(0, s) = s$. Thus

$$g(z, s) = -2s + 2h(z, s).$$

That is, we take $g(z, w) = -2w + 2h(z, w)$ and we note $g(0, w) = -2w + 2h(0, w) = 0$. Thus $g_w(0, 0) = 0$ and $g(z, w)$ begins with second-order terms since we also have

$$g_z(0, 0) = 2h_z(0, 0) = -2iF_z(0, 0, 0) = 0.$$

M now has the form

(2)
$$v = F(z, \bar{z}, u) \quad \text{with } F(0, 0, u) = 0, \quad F_z(0, 0, 0) = 0, \quad F(z, 0, u) = 0.$$

Finally we turn to $F_{k1}(u)$. Until here, we have not used that M is non-degenerate. Now we use this by assuming $F_{11}(u) \neq 0$ near the origin. We write

$$(3) \quad F(z, \bar{z}, u) = F_{11}(u)|z|^2 + zA(\bar{z}, u) + \bar{z}\bar{A}(z, u) + \sum_{\substack{k \geq 2 \\ j \geq 2}} F_{kj}(u)z^k\bar{z}^j$$

$$= F_{11}(u)\left(z + \frac{\bar{A}(z, u)}{F_{11}(u)}\right)\left(\bar{z} + \frac{A(\bar{z}, u)}{F_{11}(u)}\right) + \sum_{\substack{k \geq 2 \\ j \geq 2}} G_{kj}(u)z^k\bar{z}^j.$$

Note that $A(z, u)$ is of second order in z. Set

$$f(z, w) = \bar{A}(z, w)/F_{11}(w)$$

and consider the biholomorphism

$$z^* = z + f(z, w),$$
$$w^* = w.$$

Thus $F^*(z^*, \bar{z}^*, u^*) = F(z, \bar{z}, u)$. Since

$$f(z, w) = f(z, u) + if_u(z, u)F(z, \bar{z}, u) + \cdots$$
$$= f(z, u) + \sum_{\substack{k \geq 2 \\ j \geq 1}} H_{kj}(u)z^k\bar{z}^j,$$

we see that

$$z + \frac{\bar{A}(z, u)}{F_{11}(u)} = z^* - \sum H_{kj}(u)z^k\bar{z}^j$$

and hence from (3)

$$F^*(z^*, \bar{z}^*, u^*) = F_{11}(u^*)|z^*|^2 + \sum_{\substack{k \geq 2 \\ j \geq 2}} F_{kj}^*(u^*)z^{*k}\bar{z}^{*j}.$$

PROOF OF PROPOSITION 2. We have M of the form

$$(4) \qquad v = F_{11}(u)|z|^2 + \sum_{\substack{k \geq 2 \\ j \geq 2}} F_{kj}(u)z^k\bar{z}^j, \qquad F_{11}(0) \neq 0,$$

and we are given a real analytic section $e(t)$ of $H^{1,0}$ along the u-axis. Thus $e(t) = h(t)\frac{\partial}{\partial z}$. We seek a biholomorphism Φ preserving the form (4) and the points on the u-axis and such that $\Phi_* e(t) = \frac{\partial}{\partial z}$. After finding such a biholomorphism we show uniqueness by proving that only the identity preserves the form (4), the points of the u-axis, and the vector field $\frac{\partial}{\partial z}$ along the u-axis. It might be worthwhile to emphasize that for the first four propositions we are dealing with a fixed parametrization of the given curve γ.

Consider the biholomorphism Φ whose inverse is given by

(5)
$$z = z^* h(w^*),$$
$$w = w^*.$$

Let $G(z, \bar{z}, u)$ denote various terms of the form $\sum_{\substack{k \geq 2 \\ j \geq 2}} G_{kj}(u) z^k \bar{z}^j$. Then

$$
\begin{aligned}
v^* = \operatorname{Im} w^* = v &= F_{11}(u)|z|^2 + G(z, \bar{z}, u) \\
&= F_{11}(u^*)|h(u+iv)|^2 |z^*|^2 + G(z^*, \bar{z}^*, u^*) \\
&= F_{11}(u^*)|h(u^*)||z^*|^2 + G(z^*, \bar{z}^*, u^*).
\end{aligned}
$$

Or more simply,

(6)
$$v^* = F_{11}^*(u^*)|z^*|^2 + G(z^*, \bar{z}^*, u^*).$$

So Φ preserves (4). It clearly preserves the points of the u-axis. Also, $\Phi_* e(t) = h(t)\Phi_* \frac{\partial}{\partial z} = \frac{\partial}{\partial z}$. Thus it is the desired biholomorphism. We note for later that $F_{32}^*(u^*)$ must be the same as $F_{32}(u)$ for any transformation of the type (5) provided $w = u + iv$ with v given by (4). This is because the lowest term in w that is not $|z|^{2k}$ is already $O(|z|^5)$ and so can contribute only to the $O(|z|^7)$ terms in (6).

LEMMA 1. *Let* $z^* = \Phi(z, w)$, $w^* = \Psi(z, w)$ *map some manifold of the form*

$$v = F_{11}(u)|z|^2 + \sum_{\substack{k \geq 2 \\ j \geq 2}} F_{kj}(u) z^k \bar{z}^j$$

to another of the same form while leaving the u-axis pointwise fixed and the vector field $\frac{\partial}{\partial z}$ along the u-axis also fixed. Then the map is the identity.

PROOF. The u-axis is left pointwise fixed. This implies that $\Phi(0, u) = 0$ and $\Psi(0, u) = u$. Thus

$$
\begin{aligned}
z^* &= z\phi(z, w), \\
w^* &= w + z\psi(z, w).
\end{aligned}
$$

We have for this transformation

(7)
$$\frac{\partial}{\partial z} \longrightarrow (\phi(z, w) + z\phi_z(z, w))\frac{\partial}{\partial z^*} + (\psi(z, w) + z\psi_z(z, w))\frac{\partial}{\partial w^*}.$$

Thus if we require $\frac{\partial}{\partial z}$ to map to $\frac{\partial}{\partial z^*}$ along the u-axis we have from the first term that $\phi(0, w) = 1$ and thus

$$z^* = z + z^2 \phi_1(z, w).$$

We need not consider the second term in (7) since a different argument will yield a stronger conclusion.

If we write the manifolds as

$$v = F(z, \overline{z}, u), \qquad v^* = F^*(z^*, \overline{z}^*, u^*),$$

then substituting into

$$v^* = v + \operatorname{Im} z \psi(z, w)$$

we derive

$$
\begin{aligned}
(8) \qquad F^*(z^*, \overline{z}^*, u^*) = F(z, \overline{z}, u) &+ \frac{1}{2i}\{z\psi(z, u + iF(z, \overline{z}, u)) \\
&- \overline{z}\,\overline{\psi}(\overline{z}, u - iF(z, \overline{z}, u))\}.
\end{aligned}
$$

Now we again think of $\overline{z}$ as a separate variable and note that $\overline{z} = 0$ implies $\overline{z}^* = 0$. So if we set $\overline{z} = 0$ in (8) and use that $F^*(z^*, 0, u^*) = 0 = F(z, 0, u)$, we derive that $\psi(z, u) = 0$, and thus our transformation is of the form

$$
\begin{aligned}
(9) \qquad z^* &= z + z^2 \phi_1(z, w), \\
w^* &= w.
\end{aligned}
$$

Further (8) becomes

$$F^*(z^*, \overline{z}^*, u^*) = F(z, \overline{z}, u),$$

which we rewrite as

$$F(z, \overline{z}, u) = F_{11}^*(u)\left| z + z^2 \phi_1(z, u + iF(z, \overline{z}, u)) \right|^2 + G(z, \overline{z}, u).$$

If we now differentiate with respect to $\overline{z}$ and then set $\overline{z} = 0$, we obtain

$$F_{11}(u)z = F_{11}^*(u)(z + z^2 \phi_1(z, u)).$$

Thus $\phi_1(z, w) = 0$ and our transformation (9) is the identity.

PROOF OF PROPOSITION 3. We have just seen that given $\gamma(t)$ and $e_0(t)$ there is a unique map ϕ realizing

$$v = F_{11}(u)|z|^2 + G_{2,2}$$

with

$$
\begin{aligned}
\phi(\gamma(t)) &= (0, t), \\
\phi_* e(t) &= \frac{\partial}{\partial z}\Big|_{(0,t)},
\end{aligned}
$$

and that the unique map for $\gamma(t)$ and $h(t)e_0(t)$ realizes

$$v = F_{11}(u)|h(t)|^2 |z|^2 + G_{2,2}.$$

We draw two consequences:

1. The quantity

$$p(t) = \frac{|e_0(t)|}{\sqrt{F_{11}(t)}}$$

is independent of the choice of $e_0(t)$.

2. If $|e_1(t)| = p(t)$ then the unique map for $\gamma(t)$ and $e_1(t)$ realizes

$$v = \pm|z|^2 + G_{2,2}.$$

Note that by reversing the parametrization and mapping w to $-w$ one can reduce to the case where $F_{11}^*(u^*) = +1$. So from now on we will treat only this case.

PROOF OF PROPOSITION 4. We have shown that starting with any analytic parametrized curve $\gamma(t)$ on our nondegenerate hypersurface M there is a unique positive analytic function, call it $p(t)$, such that for each analytic section $e(t)$ of $H^{1,0}$ along γ and with $|e(t)| = p(t)$ there is a unique biholomorphism that takes

(i) $\gamma(t)$ to $(0, t)$ for all small t,
(ii) M to $\{v = |z|^2 + \sum_{\substack{k \geq 2 \\ j \geq 2}} F_{kj}(u)z^k\overline{z}^j\}$,

(iii) $e(t)$ to $\frac{\partial}{\partial z}$.

We now want to pick $e(t)$ in such a way that also

(iv) $F_{22}(u) = 0$.

First let $\tilde{e}(t)$ be any section with $|\tilde{e}(t)| = p(t)$ and let Φ be the corresponding unique biholomorphism. We claim it suffices to study maps of $\Phi(M)$.

LEMMA 2. *Given any* $\lambda_0 \in \mathbb{C}$, $|\lambda_0| = 1$, *there is a unique function* $\lambda(t)$ *with* $\lambda(0) = \lambda_0$, $|\lambda(t)| = 1$, *with the property that the unique biholomorphism taking*

(i$'$) $(0, t)$ *to* $(0, t)$ *for all small* t,
(ii$'$) $\Phi(M)$ *to another of the same form, namely*

$$v = |z|^2 + \sum_{\substack{k \geq 2 \\ j \geq 2}} F_{kj}(u)z^k\overline{z}^j,$$

(iii$'$) $\frac{\partial}{\partial z}$ *to* $\lambda(t)\frac{\partial}{\partial z}$ *along the u-axis,*

also achieves

(iv$'$) $F_{22}(u) = 0$.

It is easy to see that Proposition 4 follows from this lemma: denote the biholomorphism of the lemma by ψ and set $v(t) = p^{-1}(t)\lambda^{-1}(t)\tilde{e}(t)$. Thus $(\psi \circ \Phi)_*(p(t)v(t)) = \psi_*(\lambda^{-1}(t)\frac{\partial}{\partial z}) = \frac{\partial}{\partial z}$ and since $\psi \circ \Phi$ achieves the form given in (ii$'$), it must be the biholomorphism corresponding to $\rho(t)\nu(t)$. Of course it also achieves (iv$'$). Further, given $\gamma(t)$ (which then determines $\rho(t)$) and $\tilde{e}(t)$, the unit vector $\nu(t)$ is determined by a first-order ordinary differential equation if $\lambda(t)$ is so determined. Finally, it is easy to see that any other choice for $\tilde{e}(t)$ gives the same $\nu(t)$.

We start the proof of Lemma 2 with an exercise.

EXERCISE 2. Given an analytic function $\lambda(t)$ the only biholomorphism that satisfies (i$'$), (ii$'$), (iii$'$) is

(10)
$$z^* = \lambda(w)z,$$
$$w^* = w.$$

(Hint: Using Proposition 2 is even simpler than the direct proof.)

We want to choose λ so as to satisfy (iv$'$). Write the hypersurface in (ii$'$) as

$$v = |z|^2 + F_{22}(u)|z|^4 + G_{32} + G_{23},$$

where G_{mn} shall stand for various convergent power series of the form $\sum_{\substack{k \geq m \\ j \geq n}} g_{kj}(u)z^k\bar{z}^j$. Of course, G^*_{mn} then means a series of the form $\sum_{\substack{k \geq m \\ j \geq n}} h_{kj}(u^*)z^{*k}\bar{z}^{*j}$. From (10) evaluated at $w = u + iv(z, \bar{z}, u)$ we see that

$$z^{*k}\bar{z}^{*j} = \lambda(u)^k\overline{\lambda(u)}^j z^k\bar{z}^j + G_{k+1, j+1},$$

and so $G_{mn} = G^*_{mn}$. Note also that

$$\lambda(u + iv) = \lambda(u + i(|z|^2 + G_{22}))$$
$$= \lambda(u)\left(1 + \frac{\lambda'(u)}{\lambda(u)}i|z|^2 + G_{22}\right).$$

We will assume that $|\lambda(u)| = 1$. So

$$|\lambda(u + iv)|^2 = 1 + 2i|z|^2\frac{\lambda'(u)}{\lambda(u)} + G_{22}.$$

Now we have

$$v = v^* = |z^*|^2 + F^*_{22}(u^*)z^{*2}\bar{z}^{*2} + G_{32} + G_{23}$$
$$= (1 + 2i|z|^2\frac{\lambda'}{\lambda})|z|^2$$
$$+ F^*_{22}(u)z^2\bar{z}^2 + G_{32} + G_{23},$$

from which we see that

$$F_{22}(u) = F^*_{22}(u) + 2i\frac{\lambda'(u^*)}{\lambda(u^*)}.$$

So choosing $\lambda(u)$ to satisfy the equation

(11)
$$\lambda'(u) = -\frac{i}{2} F_{22}(u)\lambda(u)$$

yields that $F_{22}^* = 0$.

EXERCISE 3. Verify that any solution to (11) does have constant absolute value and so our assumption that $|\lambda(u)| = 1$ is justified.

This concludes the proof of Proposition 4.

PROOF OF PROPOSITION 5. We start with some $\gamma(t)$ and unit vector ν_0 at p and obtain the unique biholomorphism that takes $\gamma(t)$ to $(0, t)$, ν_0 to a positive multiple of $\frac{\partial}{\partial z}$ at $(0, 0)$, and maps M to

$$M_1 = \left\{ v = |z|^2 + \sum_{\substack{k \geq 2 \\ j \geq 2}} F_{kj}(u) z^k \bar{z}^j, \quad F_{22}(u) = 0 \right\}.$$

We look for maps that take M_1 to a hypersurface of the same form and map the u-axis to itself. From among these maps we find all that achieve $F_{33}(u) = 0$. Composing each such map with the original map of M to M_1 we end up with biholomorphisms each associated to a reparametrized curve γ and some $\tilde{\nu}_0$. So if we compose with

$$z^* = \beta z, \qquad w^* = w,$$

for the appropriate β with $|\beta| = 1$, then we conclude that the unique biholomorphism associated to the reparametrized curve γ and ν_0 achieves $F_{22}(u) = 0$ and $F_{33}(u) = 0$. It will be an easy exercise to see that this biholomorphism preserves the condition $F_{32}(u) = 0$ provided this condition is valid for the original M.

LEMMA 3. *A biholomorphism maps some manifold of the form*

(12)
$$v = |z|^2 + \sum_{\substack{k \geq 2 \\ j \geq 2}} F_{kj}(u) z^k \bar{z}^j, \quad F_{22}(u) = 0,$$

to another of the same form and maps the u-axis to itself if and only if the biholomorphism is given by

(13)
$$z^* = \alpha z (q'(w))^{1/2},$$
$$w^* = q(w),$$

where α is a constant with $|\alpha| = 1$ and $q(u)$ is real with $q'(0) > 0$.

Note that we do not require that the biholomorphism maps the u-axis pointwise to itself. The proof is based on several lemmas where we consider mappings preserving a more general form than (12).

EXERCISE 4. Verify that maps of the form (13) do preserve (12) and also preserve the condition that $F_{32} = 0$. In proving the second statement, do not assume $F_{22} = 0$.

LEMMA 4. *A biholomorphism maps some manifold of the form*

(14)
$$v = F_{11}(u)|z|^2 + \sum_{\substack{k \geq 2 \\ j \geq 2}} F_{kj}(u)z^k \bar{z}^j$$

to another of this form and maps the u-axis to itself only if the biholomorphism is of the form

(15)
$$z^* = z\phi(w), \qquad \phi(0) \neq 0,$$
$$w^* = \psi(w), \qquad \psi(u) \ real, \qquad \psi'(0) \neq 0.$$

Further, if the biholomorphism is of this form, then it maps every manifold of the form (14) *to another of the same form.*

EXERCISE 5. Verify that maps of the form (15) do preserve (14) and that if the u-axis is pointwise preserved then $\psi(w) = w$.

PROOF. A particular case of the exercise is that any map of the form

(16)
$$z_1 = z,$$
$$w_1 = \lambda(w),$$

where $\lambda(w)$ is real when w is real and $\lambda'(0) \neq 0$, preserves the form (14). So let

(17)
$$z^* = \Phi(z, w),$$
$$w^* = \psi(z, w),$$

be a biholomorphism satisfying the conditions of the lemma. Since the u-axis is preserved we must have that $\psi(0, w)$ is real when w is real. Thus we may define λ by $\lambda(\psi(0, w)) = w$ and compose (17) and (16) to obtain a map

(18)
$$z_1 = \Phi(z, w),$$
$$w_1 = \lambda(\psi(z, w)),$$

which preserves the form (14) and also preserves pointwise the u-axis. We can go further. Any map of the form

$$z_2 = z_1\phi(w_1),$$
$$w_2 = w_1,$$

also preserves the form (14) and preserves pointwise the u-axis. Note that along the u-axis

$$\frac{\partial}{\partial z} \longrightarrow \phi(u_1)\frac{\partial}{\partial z},$$

and so for an appropriate choice of $\phi(w)$ we have that the map

$$z_2 = \Phi(z, w)\phi(\lambda(\psi(z, w))),$$
$$w_2 = \lambda(\psi(z, w)),$$

preserves (14), the u-axis, and the vector field $\frac{\partial}{\partial z}$ along the u-axis. By Lemma 1, this map is the identity. Clearly, this implies $\psi(z, w)$ is only a function of w and $\Phi(z, w)$ is of the form $z\phi(w)$.

PROOF OF LEMMA 3. Thanks to Lemma 4 we may start with

$$z^* = z\phi(w),$$
$$w^* = \psi(w).$$

Now we use the fact that the transformation preserves not only the form (14) but even this form with $F_{11} = 1$ and $F_{22} = 0$. Thus

(19) $$\operatorname{Im} \psi(w) = |z\phi(w)|^2 + O(|z|^5),$$

when $w = u + iv = u + i(|z|^2 + O(|z|^5))$.

For the left-hand side of (19) we have

(20)
$$\operatorname{Im} \psi(w) = \operatorname{Im}\{\psi(u) + \psi'(u)i|z|^2 + O(|z|^5)\}$$
$$= \psi'(u)|z|^2 + O(|z|^5),$$

since $\psi(u)$ is real. For the right-hand side of (19) we have

(21)
$$|z\phi(w)|^2 = |z|^2|\phi(u) + \phi'(u)i|z|^2|^2 + O(|z|^6)$$
$$= |z|^2|\phi(u)|^2 - 2\operatorname{Im}\{\overline{\phi(u)}\phi'(u)\}|z|^4 + O(|z|^6).$$

If we equate (20) and (21) we obtain

(22) $$\psi'(u) = |\phi(u)|^2$$

and

$$\operatorname{Im}\{\overline{\phi(u)}\phi'(u)\} = 0.$$

Finally, as we now show, this implies

$$\phi(w) = \alpha(\psi'(w))^{1/2},$$

for some constant α with $|\alpha| = 1$. For we may certainly write $\phi(w) = \alpha(w)\psi'(w)^{1/2}$ for some holomorphic function $\alpha(w)$ with $|\alpha(u)| = 1$. But

$$\operatorname{Im}\{\overline{\phi(u)}\phi'(u)\} = \psi'(u)\operatorname{Im}\{\overline{\alpha}(u)\alpha'(u)\},$$

and so $\operatorname{Im}\{\overline{\alpha}(u)\alpha'(u)\} = 0$. But then, $|\alpha(u)| = 1$ implies α must be a constant.

This concludes the proof of Lemma 3.

So we now seek $q(w)$ for which the transformation

(23)
$$z^* = \alpha z(q'(w))^{1/2},$$
$$w^* = q(w)$$

makes $F_{33}^*(u^*) = 0$. In doing this, we will use that F_{22} and F_{22}^* are already known to be zero. In (23), α is a constant of norm equal to one and $q(u)$ is real with $q(0) = 0$ and $q'(0) > 0$. To simplify notation we again use $\phi(w)$ for $\alpha(q'(w))^{1/2}$ and write for our transformation

(24) $$z^* = z\phi(w), \qquad w^* = q(w).$$

In the usual fashion, we compute the $|z|^6$ coefficient in both $v^* = \operatorname{Im} q(w)$ and $v^* = F^*(z^*, \overline{z}^*, u^*)$. Setting these equal will give an equation containing $F_{33}^*(u^*)$ and hence a condition guaranteeing that F_{33}^* is zero. We take our hypersurfaces to be

$$v = F \quad \text{and} \quad v^* = F^*.$$

Thus

$$v^* = F^*\big(z\phi(u + iF),\ \operatorname{Re} q(u + iF)\big),$$

and

$$(25) \qquad v^* = q'(u)F - \frac{1}{6}q'''(u)F^3 + O(|F|^4).$$

Since F is of the form

$$F = |z|^2 + \text{higher-order terms},$$

we see that the coefficient of $|z|^6$ in (25) must be

$$q'(u)F_{33}(u) - \frac{1}{6}q'''(u).$$

Next we assume F^* is of the form

$$(26) \qquad F^*(z^*, \overline{z}^*, u^*) = |z^*|^2 + \sum_{\substack{k \geq 2 \\ j \geq 2}} F_{kj}^*(u^*)z^{*k}\overline{z}^{*j}, \qquad F_{22}^* = 0,$$

and for z^* and u^* we substitute the functions derived from (24). We start with

$$(27) \qquad \begin{aligned} |z^*|^2 &= |z|^2|\phi(u + iF)|^2 \\ &= |z|^2|\phi(u) + \phi'(u)iF - \frac{1}{2}\phi''(u)F^2|^2 + O(|z|^8) \end{aligned}$$

and now use that F is of the form

$$F = |z|^2 + O(|z|^5).$$

Thus the coefficient of $|z|^6$ in (27) is $-\operatorname{Re}(\overline{\phi}\,\phi'') + |\phi'|^2$. The only term in $\sum F_{kj}^*(u^*)z^{*k}\overline{z}^{*j}$ that can contribute $|z|^6$ is $F_{33}^*(q(u))|\phi(u)|^6$. Thus, equating the $|z|^6$ coefficients in equations (25) and (26) yields

$$q'(u)F_{33}(u) - \frac{1}{6}q'''(u) = |\phi'|^2 - \operatorname{Re}(\overline{\phi}(u)\phi''(u)) + F_{33}^*(q(u))|\phi(u)|^6.$$

Now the condition that F_{33}^* be zero yields

$$q'(u)F_{33}(u) - \frac{1}{6}q'''(u) + \operatorname{Re}\overline{\phi}(u)\phi''(u) - |\phi'|^2 = 0.$$

So, since $\phi = \alpha(q')^{1/2}$,

$$(28) \qquad \frac{1}{3}q''' - \frac{1}{2}\frac{(q'')^2}{q'} + q'F_{33} = 0.$$

When this equation is satisfied, the transformation (23) realizes $F_{33}^* = 0$.

LEMMA 5. *If $q(t)$ satisfies*

$$\frac{1}{3}q''' - \frac{1}{2}\frac{(q'')^2}{q'} + q'F_{33}(t) = 0, \qquad q(0) = 0,$$

then all other solutions are given by

$$s(t) = \frac{q(t)}{aq(t) + b},$$

for constants a and b.

EXERCISE 6. Either show directly that $s(t)$ is also a solution and hence any solution is of this form or recall the Schwarzian derivative and its properties.

PROOF OF PROPOSITION 6. We start with a nondegenerate hypersurface M and a point $p \in M$. Then we find a local biholomorphism of (M, p) to $(M_1, 0)$ with M_1 of the form

(29) $$v = |z|^2 + G_{22}.$$

It is clearly sufficient to show that Proposition 6 holds for $(M_1, 0)$ in place of (M, p). That is, we may assume that M itself satisfies (29). So we fix a transverse direction in M at 0 and show that there is a unique curve γ through 0 tangent to that direction with the property that there exists some biholomorphism taking M to a hypersurface M^* of the same form but also with $F_{32}(u) = 0$, and taking γ to the u-axis. We start with any transverse curve γ and parametrize it, near the origin, by using the u-axis. So

$$\gamma(\xi) = (p(\xi), q(\xi)) = (p(\xi), \xi + iv(p(\xi), \overline{p}(\xi), \xi)),$$

with v given by (29). The curve γ has the fixed direction at the origin, so we consider $p'(0)$ as given. We seek to show that the existence of a biholomorphism from M to some M^* with $F^*_{32} = 0$ uniquely specifies γ.

The inverse of any biholomorphism that takes γ to the u-axis and preserves the above parametrization may be written as

(30)
$$z = p(w^*) + \sum_{j=1}^{\infty} c_j(w^*)z^{*j},$$

$$w = q(w^*) + \sum_{j=1}^{\infty} \psi_j(w^*)z^{*j}.$$

Because

$$q(w^*) = w^* + iv(p(w^*), \overline{p(w^*)}, w^*),$$

we see that different curves γ correspond to different functions $p(w^*)$. Our first task is to see how the other coefficients in (30) depend on the choice of the function p. With this in mind, we subject (30) to a normalization which will make it unique. First, of course, we assume that the image M^* also is of the form (29). Next we fix some choice of a $H^{1,0}$ vector field on M near

γ. Indeed we may as well, in a neighborhood of the origin in $\mathbb{C}^2$, take the vector field

$$L = \frac{\partial}{\partial z} + \beta(z, w, \overline{z}, \overline{w})\frac{\partial}{\partial w},$$

where β is determined by

$$L\left(\frac{w - \overline{w}}{2i} - \left(|z|^2 + G_{22}\left(z, \overline{z}, \frac{w + \overline{w}}{2}\right)\right)\right) = 0.$$

So we may write β, along γ, as

$$\beta(p) = \beta(p, \overline{p}, q, \overline{q}) = \tilde{\beta}(p, \overline{p}, \xi).$$

If we require that the map (30) takes L along γ to a real multiple of $\partial/\partial z^*$ then the map (30) is uniquely specified, according to Propositions 1, 2, and 3. In this case,

$$c_1(w^*) = 1 \quad \text{and} \quad \psi_1(w^*) = \beta(p(w^*)).$$

LEMMA 6. *If* (30) *takes one manifold of the form* (29) *to another of the same form, then there are real analytic functions* f_2, f_3, g_2, g_3 *such that*

$$c_2(\xi) = f_2(p, p', c_1, c_1', \psi_1, \psi_1', \xi),$$
$$c_3(\xi) = f_3(p, p', p'', c_1, c_1', \psi_1, \psi_1', c_2, c_2', \psi_2, \psi_2', \xi),$$
$$\psi_2(\xi) = g_2(p, p', c_1, \psi_1, \xi),$$
$$\psi_3(\xi) = g_3(p, p', p'', c_1, c_1', \psi_1, \psi_1', c_2, c_2', \psi_2, \psi_2', \xi).$$

In particular, if the map is normalized by taking L *to a real multiple of* $\partial/\partial z^*$ *then there are analytic functions* h_2, h_3, k_2, k_3 *such that*

$$c_1(\xi) = 1, \qquad \psi_1(\xi) = \beta(p(\xi)),$$
$$c_2(\xi) = h_2(p, p', \xi),$$
$$c_3(\xi) = h_3(p, p', p'', \xi),$$
$$\psi_2(\xi) = k_2(p, p', \xi),$$
$$\psi_3(\xi) = k_3(p, p', p'', \xi).$$

In this lemma the argument of each function is understood to also include the conjugate of the ones listed. The point of the lemma, of course, is that the same functions work for all choices of the curve γ (i.e., for all choices of p); only the arguments are changed. To simplify the presentation somewhat, we only prove the particular case where $c_1 = 1$, $\psi_1 = \beta(p)$.

Since on M^*

$$(31) \qquad w^* = \xi + i\{|z^*|^2 + G^*(z^*, \overline{z}^*, \xi)\}, \qquad G^* = O(|z^*|^4),$$

any function $\phi(w^*)$ may be expanded,

$$\phi(w^*) = \phi(\xi) + i\phi'(\xi)\{|z^*|^2 + G^*\} - \frac{1}{2}\phi''(\xi)|z^*|^4 + O(|z^*|^6).$$

In particular, from (30),

(32) $$z = p(\xi) + S \quad \text{and} \quad u = \xi + U,$$

with

(32a) $$\begin{aligned} S = & \, ip'\{|z^*|^2 + G^*\} - \frac{1}{2}p''|z^*|^4 \\ & + z^* + (c_2 + ic_2'|z^*|^2)z^{*2} \\ & + (c_3 + ic_3'|z^*|^2)z^{*3} + O(|z^*|^6) \\ & + c_4 z^{*4} + c_5 z^{*5}, \end{aligned}$$

and

(32b) $$\begin{aligned} U = & -(v_p p' + v_{\bar{p}}\bar{p}' + v_\xi)(|z^*|^2 + G^* \\ & + \mathrm{Re}\left((\psi_1 + i\psi_1'\{|z^*|^2 + G^*\} - \frac{1}{2}\psi_1''|z^*|^4)z^* \right. \\ & + (\psi_2 + i\psi_2'|z^*|^2)z^{*2} \\ & \left. + (\psi_3 + i\psi_3'|z^*|^2)z^{*3} + \psi_4 z^{*4} + \psi_5 z^{*5} \right) \\ & + O(|z^*|^6). \end{aligned}$$

From $v = \mathrm{Im}\, w$, (29), and (30), we see that

(33) $$\begin{aligned} \frac{1}{2i}\Big\{ & w^* + iv(p(w^*), w^*) - \overline{(w^* + iv(p(w^*), w^*))} \\ & + \psi_j(w^*)z^{*j} - \overline{\psi_j(w^*)z^{*j}} \Big\} \end{aligned}$$

must be equal to

(34) $$|z|^2 + G(z, \bar{z}, u)$$

when w^* is given by (31) and z and u are given by (32).

Let $\mathscr{F}$ denote various functions of $(p, p', \bar{p}, \bar{p}', \xi)$ and let g_j denote such functions which, in addition, are bounded by

$$|g_j| \le c|p|$$

near $\xi = 0$ where c can depend on p, p', and ξ. In these functions $\mathscr{F}$ we replace ψ_1 by $\beta(p(\xi))$. We seek to determine c_2 and ψ_2. We first compute the coefficient of $z^{*2}\bar{z}^*$ in (33) and (34). We work modulo terms of order $|z^*|^4$, so we may replace w^* by $\xi + i|z^*|^2$ and $p(w^*)$ by $p(\xi) + ip'(\xi)|z^*|^2$. It is then clear that (33) contributes only a term involving ψ_1', i.e., a term of the form $\mathscr{F}$. To find the coefficient in (34) we replace S by

$$ip'|z^*|^2 + z^* + c_2 z^{*2},$$

and U by

$$-v'|z^*|^2 + \mathrm{Re}(\psi_1 z^* + i\psi_1'|z^*|^2 z^* + \psi_2 z^{*2}).$$

Thus the coefficient of $z^{*2}\bar{z}^*$ in (34) is

$$c_2 - ip' + g_1 c_2 + g_2 \psi_2 + \mathscr{F},$$

and we obtain an equation of the form

(35) $$(1 + g_1)c_2 + g_2\psi_2 = \mathscr{F}.$$

We next consider the z^{*2} coefficient.

In (33), this coefficient is $\psi_2(\xi)$ while in (34) it is $\frac{1}{2}G_u(p, \bar{p}, \xi)\psi_2 + G_z c_2 + \mathscr{F} + \bar{p}c_2$. Thus

(36) $$(1 + g_3)\psi_2 + g_4 c_2 = \mathscr{F}.$$

It is clear that we may uniquely determine c_2 and ψ_2 from (35) and (36).

We proceed in the same manner to solve for c_3 and ψ_3. First we consider the coefficient of $z^{*3}\bar{z}^*$. Thus, neglecting $|z^*|^4$ and $O(|z^*|^5)$ we have

$$w^* = \xi + i|z^*|^2 + \cdots,$$
$$p(w^*) = p(\xi) + ip'(\xi)|z^*|^2 + \cdots,$$
$$S = ip'|z^*|^2 + z^* + (c_2 + ic_2'|z^*|^2)z^{*2} + c_3 z^{*3} + \cdots,$$
$$U = g_5|z^*|^2 + \mathrm{Re}((\psi_1 + i\psi_1'|z^*|^2)z^* + (\psi_2 + i\psi_2'|z^*|^2)z^{*2}$$
$$+ (\psi_3)(z^{*3})) + \cdots.$$

Substituting these values into (33) and (34), we obtain an equation analogous to (35),

(37) $$(1 + g_6)c_3 + g_7\psi_3 = \mathscr{G}.$$

In addition to the argument of $\mathscr{F}$, $\mathscr{G}$ also depends on $c_2(\xi)$ and $\psi_2(\xi)$ and their first derivatives.

Finally we consider the coefficient of z^{*3}. Here we obtain

(38) $$(1 + g_8)\psi_3 + g_9 c_3 = \mathscr{G}.$$

From (37) and (38) we determine c_3 and ψ_3.

This concludes the proof of the lemma.

Now we may derive the equation for p''. We proceed as usual by setting (33) equal to (34) and looking at various coefficients. So we start with

$$\frac{1}{2i}\{(w^* + iv(p(w^*), w^*)) - \overline{(w^* + iv(p(w^*), w^*))}$$

(39) $$+ \psi_j(w^*)z^{*j} - \overline{\psi_j(w^*)z^{*j}}\}$$
$$= |z|^2 + G(z, \bar{z}, u)$$

where

$$w^* = \xi + i\{|z^*|^2 + G^*(z^*, \bar{z}^*, \xi)\},$$
$$z = p(\xi) + S, \qquad u = \xi + U \quad (\text{see } (32)).$$

We shall see that F_{32}^* is zero precisely when $p(\xi)$ satisfies a certain ordinary differential equation. In order to do this, we must compute the $|z^*|^4$ and $|z^*|^4 z^*$ terms in (39).

First we seek the coefficients of $|z^*|^4$ and $|z^*|^4 z^*$ in the expansion of $G(z, \bar{z}, u)$. From the expressions for S and U in (32a) and (32b) it is clear that c_3' and ψ_3' do not enter into these coefficients. Thus the highest derivative of p that may occur is the second and, as is easily seen, when these coefficients are evaluated at $\xi = 0$, then no $p''(0)$ actually occurs. So let us use $\mathcal{V}$ to denote functions of (p, p', p'', ξ) with the property that for $f \in \mathcal{V}$ we have $f(0, p'(0), p''(0), 0) = 0$. Hence in (39) we may replace G by $\mathcal{V}$ when considering the coefficients of $|z^*|^4$ and $|z^*|^4 z^*$.

Now let $\mathcal{F}$ denote functions depending on (p, p', ξ) that are zero if p or p' is zero.

With this convention we write the coefficient of $|z^*|^4$ in $|z|^2$ as

$$Q_1 = -\operatorname{Re}(\bar{p}p'') - 2\operatorname{Im}(\bar{p}p')F_{22}^* + \mathcal{F} = \mathcal{V} + \mathcal{F},$$

and the coefficient of $|z^*|^4 z^*$ in $|z|^2$ as

$$Q_2 = -\frac{1}{2}\overline{p''} + i\bar{p}p'F_{32}^* - i\overline{p'}F_{22}^* + \mathcal{F} = -\frac{1}{2}\overline{p''} + \mathcal{F},$$

where all functions are evaluated at ξ.

The corresponding coefficients in the right-hand side of (39) are

$$Q_1 + \mathcal{V} \quad \text{and} \quad Q_2 + \mathcal{V}.$$

For the left-hand side we easily see that $(2i)^{-1}(w^* - \overline{w}^*)$ contributes

$$iF_{22}^*|z^*|^4 + iF_{32}^*|z^*|^4 z^*,$$

$v(p(w^*), \dots)$ contributes only $\mathcal{F}$ and $\sum \psi_j(w^*)z^{*j}$ contributes only

$$(i\psi_1'F_{22}^* - \frac{1}{2}\psi_1'')|z^*|^4 z^*.$$

Thus from (39) we obtain

(40)
$$F_{22}^* = \mathcal{F} + \mathcal{V}$$

and

(41)
$$F_{32}^* + \psi_1'F_{22}^* + \frac{i}{4}(\psi_1'' - \overline{\psi_1'}) = -\frac{1}{2}\overline{p''} + \mathcal{F}.$$

Thus the condition under which F_{32}^* equals zero is

(42)
$$-\frac{1}{2}\overline{p''} = \mathcal{F} + \mathcal{V} - \frac{i}{4}(\psi_1'' - \overline{\psi_1'}).$$

Recall, finally, that $\psi_1 = \beta(p)$. It is easily seen that $\beta(z, w, \bar{z}, \overline{w}) = 2i\bar{z} + O(|z|^3)$. Thus $\psi_1'' = 2ip'' + \mathcal{V}$ and (42) together with its conjugate may be solved for p'':

(43)
$$p'' = \mathcal{F}(p, p', \bar{p}, \overline{p'}, \xi).$$

This is the condition that F_{32}^* be zero.

Let us be clear about what has been accomplished so far. We fix a point p on a nondegenerate hypersurface M and also fix a transverse direction at p. We have found a curve in this direction through p which may be mapped to the u-axis by a biholomorphism Φ which achieves

$$v = |z|^2 + \sum_{\substack{k \geq 2 \\ j \geq 2}} F_{kj}(u)z^k\bar{z}^j \quad \text{with } F_{32}(u) = 0.$$

We now claim that there can be no other such curve. For if there exists a second curve then, starting with $\phi(M)$, we have a hypersurface

$$v = |z|^2 + \sum_{\substack{k \geq 2 \\ j \geq 2}} F_{kj}(u)z^k\bar{z}^j, \qquad F_{32}(u) = 0,$$

which is mapped to another of the same type by a biholomorphism that takes a curve γ to the u-axis. Here γ is a curve through the origin and tangent to the u-axis. We compute F_{32}^* by parametrizing γ by the u-axis. Then $F_{32}^* = 0$ becomes equation (43) where γ is given by $\gamma(\xi) = (p(\xi), \xi + iv(p(\xi), \xi))$. But since γ is tangent to the u-axis, the initial conditions for this equation are $p(0) = 0$ and $p'(0) = 0$. But for these values, the right-hand side is zero. Thus $p \equiv 0$ and γ coincides with the u-axis. This completes the proof of Proposition 6.

Once p is specified there is a five-parameter set of maps of M into normal form. The first nonzero terms in any one of these normal forms are $F_{42}(u)$ and its conjugate, $F_{24}(u)$. When these are nonzero it is possible to impose further restrictions so that we end up with only two maps into normal form. See [CM, p. 247] and the Erratum [CMa]. This same reduction will appear in a very different approach to CR structures (see Chapters 6 and 7).

We may now easily describe all mappings of a nondegenerate hypersurface M into normal form. First we specify which point $p \in M$ should map to the origin. We then choose some $e(0)$ in $H_p^{1,0}$, of a specified length, to be mapped to $\partial/\partial z$ and some transverse vector v to be mapped to $\partial/\partial u$. There is a unique curve in the direction v that maps to the u-axis. The fact that v maps to $\partial/\partial u$ partially determines the parametrization on this curve; any real number completes this determination. Thus, locally, the set of maps to normal form is parametrized by $\mathbb{R}^3 \times S^1 \times \mathbb{R}^3 \times \mathbb{R}^1$ and, in particular, is eight-dimensional. (In Chapter 6 we construct an eight-dimensional bundle over M which may be interpreted as the "bundle of maps to normal form".) What does this tell us about the equivalence problem posed at the start of the previous chapter? If (M, p) and (M', p') are biholomorphic then the set of normal forms we obtain for (M, p) must coincide with the set we obtain for (M', p'). That is, if $(N, 0)$ and $(N', 0)$ are choices for the corresponding normal forms, then $(N', 0)$ belongs to the five-parameter family of possible normal forms for $(N, 0)$. Thus, we replace the original problem for the equivalence of (M, p) and (M', p') under the infinite-dimensional pseudo-group of local biholomorphisms taking p to p' by the simpler one for the equivalence of $(N, 0)$ and $(N', 0)$ under the five-dimensional pseudo-group of maps of $(N, 0)$ into normal form. In light of our remarks above, this equivalence problem becomes much simpler when $F_{42}(0)$ is different from zero. For then, there are only two choices for $(N, 0)$ when we impose the further restriction alluded to earlier. A necessary condition for (M, p) and (M', p') to be equivalent is that $F'_{42}(0)$ is also different from zero. Thus there are also only two possibilities for $(N', 0)$. If these coincide with the possibilities for $(N, 0)$ then (M, p) and (M', p') are equivalent; otherwise they are not.

We end this chapter with a closer look at that unique curve associated to each transverse direction.

DEFINITION. A curve γ on a nondegenerate hypersurface $M^3 \subset \mathbb{C}^2$ is called a *chain* if for each point $p \in \gamma$ there is some open set U in M and some local biholomorphism Φ such that $\Phi(U)$ has the form

$$v = |z|^2 + \sum_{\substack{k \geq 2 \\ j \geq 2}} F_{kj} \quad \text{with } F_{32}(u) = 0,$$

and

$$\Phi(\gamma \cap U) \text{ lies on the } u\text{-axis.}$$

EXERCISE 7. Let (z, u) be local coordinates centered at some point $p \in M$ and chosen, as usual, so that $\{(z, 0)\}$ is tangent to H at p. Fix some direction ν transverse to H at p. Use only the fact that each point q and each transverse direction at q determines a unique chain in order to prove that there is some function F for which the chains through p with directions close to ν are precisely the curves $(z(t), t)$ where $z(t)$ satisfies the equation

$$z'' = F(z, z', t).$$

As we shall see in Chapter 8 this equation for the chains is relatively complicated. In particular, F goes to infinity as the direction $(z'(t), 1)$ becomes tangent to H. (Actually, in Chapter 8 we work with a first-order system equivalent to this equation.) Consequently, it is not easy to investigate the behavior of chains in general. But there is one hypersurface on which the chains are well understood.

THEOREM 2. *On the hypersurface* $Q = \{(z, w): v = |z|^2\}$ *the chains are the intersections of* Q *with complex lines.*

PROOF. The line $\mathcal{L}_1 = \{(0, w)\}$ intersects Q in the u-axis which certainly is a chain. Let $\mathcal{L}_2$ be any other complex line that intersects Q in a curve. We know from the first corollary to Theorem 2.5 that there is an automorphism of Q that takes $\mathcal{L}_2 \cap Q$ to the u-axis. Thus $\mathcal{L}_2 \cap Q$ is also a chain. We obtain all chains in this way since each direction transverse to H generates a complex line that intersects Q transversally.

REMARK. S^3 and Q are biholomorphic via a projective map. Since such a map preserves complex lines, the chains on S^3 are likewise the intersections of S^3 with complex lines.

The normal form not only provides us with chains but it also provides a projective structure on each chain. In the next chapter we will describe projective structures in some detail, but for now let us just say that a family of local diffeomorphisms $f_\alpha : O_\alpha \rightarrow \gamma$, with O_α an open subset of $\mathbb{R}^1$ and γ a curve, provides a projective parametrization of γ if for each α and β there exist constants a, b, c, d such that

$$f_\alpha(t) = f_\beta \left(\frac{at + b}{ct + d} \right)$$

for all $t \in f_\alpha^{-1}(f_\alpha(O_\alpha) \cap f_\beta(O_\beta))$.

So let γ be a chain in some strictly pseudoconvex hypersurface M in $\mathbb{C}^2$ and let Φ be a local biholomorphism such that $\Phi(M)$ is in normal form with $\Phi(\gamma)$ being the u-axis. Let f be the map of the u-axis into γ given by Φ^{-1}. Thus f is a parametrization of γ. We claim that as Φ varies through all maps of M to normal form, the family of associated maps f forms a projective parametrization. To prove this we consider a second map Ψ such that $\Psi(M)$ is in normal form and $\Psi(\gamma)$ is the u-axis. We let $g = \Psi^{-1} \mid_{u\text{-axis}}$, and show that $f(t) = g\left(\frac{at+b}{ct+d}\right)$ for some constants a, b, c, and d. So for any real constant A let Λ be the biholomorphism

$$z^* = z, \qquad w^* = A + w.$$

Λ preserves the normal form in the sense that if N is in normal form then so is $\Lambda(N)$. In particular, for $A = -\Psi(f(0))$, the biholomorphism $\Lambda \circ \Psi \circ \Phi^{-1}$ takes the normal form $\Phi(M)$ into another normal form and takes the origin to itself. By earlier results, especially equation (28) and Lemma 5, $\Lambda \circ \Psi \circ \Phi^{-1}$ is of the form

$$z^* = \alpha z (q'(w))^{1/2},$$
$$w^* = q(w),$$

with $q(t) = \frac{t}{ct+d}$ for some constants c and d. Thus

$$\Lambda \circ \Psi(f(t)) = \frac{t}{ct + d}$$

and

$$\Psi(f(t)) = \frac{t}{ct + d} - A = \frac{at + b}{ct + d}.$$

So

$$f(t) = \Psi^{-1}\left(\frac{at+b}{ct+d}\right) = g\left(\frac{at+b}{ct+d}\right)$$

and we are done.

CHAPTER 5

Background for Cartan's Work

Cartan's study of CR structures is an example of his general approach to problems of differential geometry. This approach was extremely fruitful and extended to Riemannian and other differential geometries the relation to groups which was emphasized by Klein in the case of the classical geometries. This chapter provides an introduction to Cartan's general methods by considering first the most intuitive differential geometry, that of two-dimensional Riemannian geometry; second, the Frobenius Theorem and its relation to Lie groups; and third, Maurer-Cartan connections and, more generally, fibre bundles with connections. As examples we discuss in detail certain homogeneous spaces of $SU(2, 1)$ and of $PGL(2)$.

§1. A simple equivalence problem. We start with a simple equivalence problem taken from the expository paper [Ca3], namely the local equivalence problem for two-dimensional Riemannian manifolds. So let M and $\widetilde{M}$ be two such manifolds. When are they locally isometric? This is a question about partial differential equations, so we first formulate the relevant equations. We shall do so in the language of differential forms. On some open set U of M, let ω_1 and ω_2 be one-forms which are orthonormal with respect to the given metric. (If you prefer starting with vectors, take e_1 and e_2 to be orthonormal vectors and then let ω_1 and ω_2 give the dual basis.) Let $\widetilde{\omega}_1$ and $\widetilde{\omega}_2$ be an orthonormal basis on some open set $\widetilde{U}$ of $\widetilde{M}$. The fact that $\Psi: U \to \widetilde{U}$ is an isometry does not mean that $\Psi^*\widetilde{\omega}_i = \omega_i$, $i = 1, 2$; it only means that

(1)
$$\Psi^*\widetilde{\omega}_1 = (\cos \mu)\omega_1 + (\sin \mu)\omega_2,$$
$$\Psi^*\widetilde{\omega}_2 = -(\sin \mu)\omega_1 + (\cos \mu)\omega_2,$$

where μ is some function on M. This is a system of four first-order partial differential equations in the three unknowns ψ_1, ψ_2, μ. Here we think of coordinates on U and $\widetilde{U}$ and take ψ_1 and ψ_2 to be the components of Ψ. Now these equations are in quite an awkward form to analyze. The basic idea in Cartan's approach to equivalence problems is to admit μ as a new independent variable rather than an unknown in order to obtain simpler equations. Of course if we do this we will also need to find another equation.

89

LEMMA 1. *Let ω_1 and ω_2 be arbitrary forms on U with $\omega_1 \wedge \omega_2 \neq 0$. Let Ω_1 and Ω_2 be the forms on $U \times S^1$ given by*

(2)
$$\Omega_1 = (\cos \theta)\omega_1 + (\sin \theta)\omega_2,$$
$$\Omega_2 = -(\sin \theta)\omega_1 + (\cos \theta)\omega_2.$$

There is a unique form Ω_3 on $U \times S^1$ satisfying

(3)
$$d\Omega_1 = \Omega_3\Omega_2, \qquad d\Omega_2 = -\Omega_3\Omega_1.$$

EXERCISE 1. Prove the lemma and show that equations (3) imply that

$$d\Omega_3 = -K\Omega_1\Omega_2$$

where K is independent of θ. Show that K is the classical Gaussian curvature. (Hint: Seek Ω_3 of the form $d\theta + A\Omega_1 + B\Omega_2$.)

REMARK. The reason for choosing equations (3) will become clear when we discuss the structure equations of a Lie group.

Let Ψ be isometric, i.e., let Ψ satisfy (1) for some function μ. Thinking of this function μ as fixed, there exists a unique map $\Phi: U \times S^1 \to \tilde{U} \times S^1$ such that

$$\Phi^*(\tilde{\Omega}_j) = \Omega_j, \quad j = 1, 2.$$

Indeed, Φ is given by $\Phi(p, \theta) = (\Psi(p), \theta - \mu(p))$. Using that Ω_3 is characterized by (3), it is easy to show that $\Phi^*(\tilde{\Omega}_3) = \Omega_3$. Thus if there is a solution to (1) then there is a solution to

(4)
$$\Phi^*(\tilde{\Omega}_j) = \Omega_j, \quad j = 1, 2, 3.$$

This represents a system of nine first-order equations for three unknowns where each unknown is a function of three variables. The form of this system is simpler than that of (1), and we will be able to apply the Frobenius theorem to (4) in order to prove a fundamental result about surfaces of constant curvature.

First we need to establish that (4) implies (1).

EXERCISE 2. Let $\Phi: U \times S^1 \to \tilde{U} \times S^1$ satisfy (4). Show that Φ is of the form $\Phi(p, \theta) = (\Psi(p), \theta - \mu(p))$ where $\Psi: U \to \tilde{U}$ satisfies (1).

So (1) has a solution if and only if (4) has a solution.

THEOREM 5.1. *Any Riemannian manifold of dimension two with constant Gaussian curvature admits a local three-dimensional transitive group of isometries. Any two such manifolds with the same constant Gaussian curvature are locally isometric.*

REMARK. Analogous results hold for all dimensions.

PROOF. The proof is based on a simple application of the Frobenius Theorem on integrable systems of differential equations. For the reader's convenience we have placed after this proof a review of the Frobenius Theorem in several of its guises.

As before we have that Ω_j, $j = 1, 2, 3$, are one-forms on $V = U \times S^1$ and $\tilde{\Omega}_j$ are one-forms on $\tilde{V} = U \times S^1$. Let $\Psi_j = \Omega_j - \tilde{\Omega}_j$ be one-forms on $V \times \tilde{V}$; that is, on an open set in $\mathbb{R}^6$. We always have $d\Psi_1 = \Omega_3\Omega_2 - \tilde{\Omega}_3\tilde{\Omega}_2 = \Omega_3(\Omega_2 - \tilde{\Omega}_2) + (\Omega_3 - \tilde{\Omega}_3)\tilde{\Omega}_2 = \Omega_3\Psi_2 + \Psi_3\tilde{\Omega}_2$ and similarly for $d\Psi_2$. The same type of factoring works for $d\Psi_3$ when K is the same constant for V and $\tilde{V}$. Then we have

$$d\Psi_1 = \Omega_3\Psi_2 - \tilde{\Omega}_2\Psi_3,$$
$$d\Psi_2 = \Omega_3\Psi_1 - \tilde{\Omega}_1\Psi_3,$$
$$d\Psi_3 = -K\Omega_1\Psi_2 + K\tilde{\Omega}_2\Psi_1.$$

Thus $\{\Psi_1, \Psi_2, \Psi_3\}$ generate a closed differential ideal and we may apply the Frobenius Theorem. Thus, through each point $q = (p, \theta, \tilde{p}, \tilde{\theta})$ there is a three-dimensional manifold N on which each $\{\Psi_j\}$ restricts to zero. First we claim that N is a graph over V near q. To see this let $\pi_1 : V \times \tilde{V} \to V$ be the projection onto the first factor and π_2 the projection onto the second factor. Let $X \in TN_q$ with $\pi_{1*}X = 0$. If this implies $\pi_{2*}X = 0$ then N is a graph over V. Since $\pi_{1*}X = 0$, we must have that $\Omega_j(X) = 0$. But, since $\Omega_j(X) = \tilde{\Omega}_j(X)$, this implies $\tilde{\Omega}_j(X) = 0$, for $j = 1, 2, 3$. But then $\pi_{2*}X = 0$. So we have that N is of the form $\{(p, \theta, \Phi(p, \theta))\}$ where $\Phi : V \to \tilde{V}$. Since $\Psi_j = 0$ we have that $\Phi^*(\tilde{\Omega}_j) = \Omega_j$. But as we have seen, this implies that $\Phi = (\Psi(p), \theta - \mu(p))$, where $\Psi : U \to \tilde{U}$ is an isometry.

This was done in the neighborhood of any point $q = (p, \theta, \tilde{p}, \tilde{\theta})$. So there is a one-parameter family of isometries $U \to \tilde{U}$ that take a given point p to a given point $\tilde{p}$. Finally, let $M = \tilde{M}$ and let $\Psi_{(p, \theta)}$ denote the local isometry of M to itself determined by N_q with $q = (0, 0, p, \theta)$. Let $G = \{\Psi_{(p, \theta)}\}$. Then G is locally a three-dimensional group.

The Frobenius Theorem can be stated in at least three different ways. The simplest is as a result about overdetermined partial differential equations. The most useful in the type of geometric problems we study here is as a result on ideals generated by one-forms. And the most useful in many other types of geometric problems is as a result on vector fields.

1. Consider the system of equations

$$\frac{\partial u^\sigma}{\partial x_j} = f_j^\sigma(x, u),$$
$$u^\sigma(x_0) = p_0^\sigma,$$

for $x \in \mathbb{R}^n$, $\sigma = 1, \ldots, s$, (x_0, p_0^σ) a given point in $\mathbb{R}^{n+s}$ and f of class C^1 in a neighborhood of some point (x^*, p^*).

THEOREM. *This system has a solution for each* (x_0, p_0) *near* (x^*, p^*) *if and only if* f *satisfies, in a neighborhood of* (x^*, p^*), *the condition*

$$\frac{\partial f_j^\sigma}{\partial x_k} + \frac{\partial f_j^\sigma}{\partial u^a} f_k^a \quad \text{is symmetric in } j \text{ and } k.$$

2. Let $\omega_1, \ldots, \omega_n$ be independent 1-forms defined near some point $p \in \mathbb{R}^{n+m}$. A submanifold M^m is said to be an integral submanifold if each ω_j is zero when restricted to M. A form ω (of arbitrary degree) is said to satisfy the condition

$$\omega \equiv 0 \qquad \text{mod } \{\omega_1, \ldots, \omega_n\}$$

if

$$\omega \wedge \omega_1 \wedge \cdots \wedge \omega_n = 0.$$

THEOREM. *There is some neighborhood of* p *foliated by integral submanifolds if and only if*

$$d\omega_j \equiv 0 \quad \text{mod } \{\omega_1, \ldots, \omega_n\} \quad \text{for } j = 1, \ldots, n.$$

This condition can be restated: The ideal generated by $\{\omega_1, \ldots, \omega_n\}$ is a closed differential ideal.

3. Let $X_1, \ldots, X_m$ be independent vector fields defined near some point $p \in \mathbb{R}^{m+n}$. A submanifold M^m is said to be an *integral submanifold* if its tangent space at each point is spanned by $\{X_1, \ldots, X_m\}$.

THEOREM. *There is some neighborhood of* p *foliated by integral submanifolds if and only if at each point near* p

$$[X_j, X_k] \text{ is in the linear span of } \{X_1, \ldots, X_m\}.$$

This conclusion can also be restated: The m-plane distribution spanned by $\{X_1, \ldots, X_m\}$ is closed under taking brackets.

REMARK. If in 1 the functions $f_j^a(x, u)$ are C^∞ (or C^ω) then the solution $u(x)$ is also C^∞ (or C^ω). Similar results hold for 2 and 3.

EXERCISE 3. (1) Show these three-forms of the Frobenius theorem are equivalent. (See for instance [Sp, vol. I], pp. 6–11, 6–19, 6–28 (problem 7), and 7–23).

(2) Let $\omega_1, \ldots, \omega_n$ and $\theta_1, \ldots, \theta_n$ be one-forms on M^n with $\omega_1 \wedge \cdots \wedge \omega_n \neq 0$ in a neighborhood of some point p and $\theta_1 \wedge \cdots \wedge \theta_n \neq 0$ in a neighborhood of some point q. Show there is at most one map Φ that takes p to q and satisfies $\Phi^*(\theta_j) = \omega_j$, $j = 1, \ldots, n$, near p. (Hint: Introduce local coordinates and use the fundamental uniqueness theorem of ordinary differential equations.) Note that the assumption $\omega_1 \wedge \cdots \wedge \omega_n \neq 0$ is superfluous.

(3) Again let $\omega_1, \ldots, \omega_n$ and $\theta_1, \ldots, \theta_n$ be one-forms near p and q respectively. Assume

$$d\omega_k = C_{j\ell}^k \omega_j \wedge \omega_\ell, \quad k = 1, \ldots, n,$$

and

$$d\theta_k = C^k_{j\ell} \theta_j \wedge \theta_\ell, \quad k = 1, \ldots, n,$$

for the same *constants* $C^k_{j\ell}$. Show that there is a unique map Ψ in a neighborhood of p with $\Psi(p) = q$ and $\Psi^*(\theta_k) = \omega_k$, $k = 1, \ldots, n$.

§2. Maurer-Cartan connections. As we have seen in the isometries of two-dimensional manifolds, Cartan's treatment of equivalence problems has as one ingredient the Frobenius Theorem. A second ingredient is the natural connection on a Lie group. In the example this appears only implicitly—$M \times S^1$ has a local group structure. In more complicated work the group is more prominent. Recall that a Lie group is nothing more than a smooth manifold M together with a diffeomorphism $L_p : M \to M$ for each point p such that the operation $pq \equiv L_p q$ defines a group. A local Lie group structure means that L_p is defined only on some open set with L_q being the identity for some q in the set and the other axioms for a group being valid whenever they make sense.

On any Lie group there is a natural connection which takes its values in the Lie algebra. But before proceeding with this, let us review the general concept of a connection. We start with the idea that a connection should provide a way of "connecting" the structure of the manifold at its various points. So at the most naive level, we might expect that a connection relates some information in a neighborhood of any one point p of a manifold M to information in a neighborhood of any other point q. We work with infinitesimal neighborhoods and take the information to be the tangent space at the points. That is, a connection should be an isomorphism of $T_p M$ onto $T_q M$. At only a slightly less naive level, it is seen that the connecting isomorphism must depend on the curve chosen to "connect" the points p and q. That is, given points p and q of M and a curve Γ from p to q, Γ should determine an isomorphism $T_p M \to T_q M$. We use the infinitesimal version of this and replace the curve by a tangent vector. A connection at each p assigns to each $X \in T_p M$ an "infinitesimal" mapping of $T_p M$ to itself. This mapping is easiest to explain using the frame bundle for M.

A frame at a point $x \in M$ is an ordered basis $X_1, X_2, \ldots, X_n$ for $T_x M$. The fiber at the point x of the frame bundle $F(M)$ is the set of all choices of frames at x. $F(M)$ is a principal fiber bundle over M with fiber group $G = \mathrm{GL}(n)$. G acts on the right by

$$g = (a^j_i), \quad \{X_1, X_2, \ldots, X_n\} g = \{Y_1, Y_2, \ldots, Y_n\},$$

with

$$Y_i = a^j_i X_j.$$

We are using the notation of [KN]; see this reference also for more details, formal definitions, etc. We also need to work with the Lie algebra $\mathfrak{G}$ of G. The Lie algebra of any Lie group is the set of its left-invariant vector fields and can be identified with the tangent space at the identity.

We have the usual picture

$$G - F(M)$$

$$\downarrow$$

$$M$$

As we have indicated, a connection should assign to each $X \in T_x M$ some motion in the fiber. So let $\gamma(t)$ be a curve in M and let $f(t)$ be a choice of frames on $T_{\gamma(t)} M$. A connection should compare these frames to some standard frame. That is, we might think of a connection as a curve in the group, $g(t)$, depending on $\gamma(t)$ and $f(t)$ that tells us to "connect" $(\gamma(0), f(0))$ to $(\gamma(t), f(t)g(t))$. (Remember that G acts on the right.) Infinitesimally, $g(t)$ corresponds to some element $\mathfrak{g}$ of the Lie algebra $\mathfrak{G}$ and $(\gamma(t), f(t))$ to some $U \in T_p F$, $p = (\gamma(0), f(0))$. What properties should the map $U \to \mathfrak{g}$ have? The frame $(f(t))h$ should "connect" to the frame $f(t)g(t)h$. So we should take $f(t)h g_1 = f(t)g(t)h$. That is,

$$g_1(t) = h^{-1}g(t)h,$$

and so

$$\mathfrak{g}_1 = L_{h^{-1}}R_h\mathfrak{g}.$$

Since $\mathfrak{g}$ is left-invariant, this is just

$$\mathfrak{g}_1 = \mathrm{ad}_{h^{-1}}\mathfrak{g}.$$

Here ad_a is the adjoint action of a Lie group on its Lie algebra. It is induced by the map $g \to a g a^{-1}$. Further, if the curve is vertical, i.e., $(\gamma(0), f(t))$, then the connection should be given by $(\gamma(0), f(0)a(t))$. Thus the element $A = da(t)/dt$ of $\mathfrak{G}$ is associated by the connection to the fundamental vector field $A^* = d(f(0)a(t))/dt$ given by the right action of G on itself. Note that A^* is a left-invariant vector field.

We could now formalize what we mean by a connection on the frame bundle. But the same considerations hold more generally. So we give the definition of a connection on a principal fiber bundle. Let G be a Lie group, $\mathfrak{G}$ its Lie algebra, and P a principal fiber bundle with fiber group G. To each $A \in \mathfrak{G}$ we may associate the vertical vector field A^* induced by the action of G on P.

DEFINITION. A *connection* is a $\mathfrak{G}$-valued 1-form

$$\omega: TP \to \mathfrak{G}$$

that satisfies

$$R_{g*}\omega_{ug} = \mathrm{ad}(g^{-1})\omega_u$$

and

$$\omega(A^*) = A.$$

There is a natural connection on the Lie group itself (thought of as a fiber bundle over a point) which relies on the identification of the Lie algebra with the left-invariant vector fields on G. That is, we just associate to each $X \in TG$ the left-invariant vector field that it generates,

$$X \to \{V = L_{g*}X : g \in G\}.$$

This natural connection is called the Maurer-Cartan connection of the group. It can in fact be defined even more simply.

DEFINITION. The *Maurer-Cartan connection* on a group is given by the form

$$\omega(A) = A, \quad \text{for} \quad A \in \mathfrak{G}.$$

It is easily seen that this is a connection in the sense of the previous definition. It is also easily seen that ω is a left-invariant differential form (with values in the Lie algebra $\mathfrak{G}$). Let $E_1, E_2, \ldots, E_n$ be a basis for the Lie algebra and write

$$\omega = \omega^i E_i,$$

where ω^i is a real valued 1-form. It follows that each ω^i is also left-invariant and that $\{\omega_1, \omega_2, \ldots, \omega_n\}$ is linearly independent. Thus

$$d\omega^i = \sum_{1 \le j < k \le n} C^i_{jk} \omega^j \wedge \omega^k,$$

where C^i_{ji} are constants. This is the *structural equation of the group*.

Thus on any Lie group there is a natural connection and this connection has constant coefficients in its structural equation. We can also start with such a connection and manufacture the group.

EXERCISE 4. Let $\omega^1, \ldots, \omega^n$ be one-forms near $p \in M^n$. Assume $\omega^1 \wedge \cdots \wedge \omega^n \ne 0$ and $d\omega^i = C^i_{jk} \omega^j \wedge \omega^k$ where C^i_{jk} are constants. Find a local Lie group structure such that

$$L^*_q\left(\omega^k|_r\right) = \omega^k|_{q^{-1}r}, \quad k = 1, \ldots, n,$$

for all q and r near p. (Hint: Consider the unique map f_q that satisfies $f^*(\omega) = \omega$ and $f(p) = q$.)

The most important example of a Lie group for us is $G = GL(n)$ in its usual representation by matrices. It is easy to find the Maurer-Cartan form and to compute its structural equation; however, the notation is somewhat awkward. Let e_{ij} be the matrix that has zero everywhere except for a one in the i th row and j th column. Any element of G can be written as $g = x_{ij}e_{ij}$ and so $\{x_{ij}\}$ serve as coordinates for G. The group operation is given by matrix multiplication

$$L_h g = h_{ik} g_{km} e_{im}.$$

For the Lie algebra $\mathfrak{G}$, thought of as the tangent space at the identity, we may take as a basis the vectors

$$E_{ij} = \frac{\partial}{\partial x_{ij}}$$

evaluated at $x_{ij} = \delta_{ij}$. This vector E_{ij} may also be considered as a matrix, namely, the matrix all of whose components are zero, except for a one in the (i, j)-th position. Thus when this matrix is thought of as an element of G we denote it by e_{ij}, and when it is to be an element of $\mathfrak{G}$ we denote it by E_{ij}. So starting from

$$g = x_{ij} e_{ij},$$

we obtain the $\mathfrak{G}$-valued 1-form

$$dg = dx_{ij} E_{ij},$$

and using matrix multiplication

$$(5) \qquad\qquad g^{-1} dg = (g^{-1})_{ab} dx_{bc} E_{ac}.$$

We will use this expression below.

For $\mathfrak{G}$ thought of as left-invariant vector fields, we may take as a basis the vector fields which, at the identity, agree with E_{ij}. Thus

$$(6) \qquad\qquad A_{ij}\Big|_h = (L_h)^* E_{ij} = x_{Ii} \frac{\partial}{\partial e_{Ij}} = x_{Ii} E_{Ij}.$$

There is a more concrete way of doing this by focusing on the fact that we are working with matrices. We identify $GL(n)$ with an open subset of $\mathbb{R}^N$, $N = n^2$. At each point of G, the tangent space at that point can be identified with $\mathbb{R}^N$, that is, with the set $M(n)$ of $n \times n$ matrices. Let B be some element of $M(n)$. Think of B as being in $T_c G$ at some point $c \in G$. Let h be some other element of G and let C and H be the matrices identified with c and h under the identification of G with $\mathbb{R}^N$. We expect that $L_{h*} B = HB$. This is indeed so: let $g = (x_{ij})$, $B = (b_{ij})$, and $h = H = (g_{ij})$. Then

$$L_h g = hg = h_{ij} g_{jk}$$

and

$$L_{h*} B = L_{h*}\left(b_{ab} \frac{\partial}{\partial x_{ab}}\right) = b_{ab} \frac{\partial}{\partial x_{ab}} (h_{ij} x_{jk}) \frac{\partial}{\partial x_{ik}} = b_{jk} h_{ij} \frac{\partial}{\partial x_{ik}} = HB.$$

In particular, the left-invariant vector field determined by B has at the identity of G the value $C^{-1}B$. This is consistent with (6). For if we let $B = A_{ij}$ evaluated at $C \in G$ then

$$B = c_{ni} \frac{\partial}{\partial x_{nj}}, \quad \text{and so} \quad b_{ad} = \begin{cases} 0 & \text{for } d \neq j \\ c_{ai} & \text{for } d = j. \end{cases}$$

Thus

$$H^{-1}B = H_{ca}^{-1} B_{ad} = \begin{cases} 0 & \text{for } b \neq j \\ \delta_{ci} & \text{for } b = j. \end{cases}$$

Next we look at the multiplication in the Lie algebra, i.e., at the Lie bracket. This can be computed as the commutator of vector fields or as the commutator of matrices. For the vector fields we obtain

$$[A_{ab}, A_{cd}] = \left[X_{Ia}\frac{\partial}{\partial x_{Ib}}, X_{kc}\frac{\partial}{\partial x_{kd}}\right] = \delta_{cb}A_{ad} - \delta_{ad}A_{cb}.$$

At the identity, this is the vector field

$$\delta_{cb}E_{ad} - \delta_{ad}E_{cb}.$$

To compute the matrix commutator $[E_{ab}, E_{cd}] = E_{ab}E_{cd} - E_{cd}E_{ab}$, we note that the (i, j) component of E_{ab} is

$$(E_{ab})_{ij} = \delta_{ai}\delta_{bj},$$

and so

$$[E_{ab}, E_{cd}] = \delta_{ai}\delta_{bc}\delta_{dk} - \delta_{ci}\delta_{ad}\delta_{bk}$$
$$= \delta_{bc}E_{ad} - \delta_{ad}E_{cb}.$$

Thus these two ways of computing the Lie algebra product do coincide.

We have seen that the Maurer-Cartan form is defined by $\omega(A) = A$. To actually see what this form is, we interpret this as $\omega(A_{ij}) = E_{ij}$. That is,

$$\omega^{ab}(h_{ni}E_{nj})E_{ab} = E_{ij}.$$

From this we see that

$$\omega(E_{Jj}) = (h^{-1})_{iJ}E_{ij},$$

which implies

$$\omega = (h^{-1})_{iJ}E_{ij}\,dx_{Jj}.$$

Comparing this with (5), we see that for matrix groups the Maurer-Cartan connection is given by

(7) $$\omega = (g^{-1})\,dg.$$

This formalism yields

$$d\omega = (-g^{-1}dg)g^{-1} \wedge dg = -g^{-1}dg \wedge g^{-1}dg.$$

That is,

(8) $$d\omega = -\omega \wedge \omega.$$

To make this clearer, let us express the left- and right-hand sides in local coordinates. We have from (7) that

$$\omega^{ac} = (g^{-1})_{ab}dx_{bc},$$

and so

$$d\omega^{ac} = -((g^{-1}dg)g^{-1})_{ab} \wedge dx_{bc} = -(g^{-1})_{am}dx_{mn} \wedge (g^{-1})_{nb}dx_{bc}.$$

For the right-hand side,

$$\omega^{an} \wedge \omega^{nc} = (g^{-1})_{am}dx_{mn} \wedge (g^{-1})_{nb}dx_{bc}.$$

So

(9) $$d\omega^{ac} = -\omega^{an} \wedge \omega^{nc}.$$

Equation (8) can also be written in terms of $[\omega, \omega]$. For any Lie algebra, let ω and θ be $\mathfrak{G}$-valued 1-forms. The 2-form $[\omega, \theta]$ acting on two tangent vectors X and Y is defined to be

(10) $$[\omega, \theta](X, Y) = [\omega(X), \theta(Y)] - [\omega(Y), \theta(X)].$$

EXERCISE 5. Let $\omega = \omega^j E_j$ and let $\theta = \theta^j E_j$ where $\{E_1, \ldots, E_n\}$ is a basis for $\mathfrak{G}$. Show that

$$[\omega, \theta] = (\omega^j \wedge \theta^j)([E_j, E_j]).$$

In particular, for $\omega = \omega^{ab} E_{ab}$, as above, we have

$$
\begin{aligned}
[\omega, \omega] &= [\omega^{ab} E_{ab}, \omega^{cd} E_{cd}] \\
&= \omega^{ab} \wedge \omega^{cd} [E_{ab}, E_{cd}] \\
&= \omega^{ab} \wedge \omega^{cd} (\delta_{cb} E_{ad} - \delta_{ad} E_{cb}) \\
&= \omega^{ab} \wedge \omega^{bd} E_{ab} - \omega^{ab} \wedge \omega^{ca} E_{cb} \\
&= \omega^{ab} \wedge \omega^{bd} E_{ad} - (-\omega^{ca} \wedge \omega^{ab}) E_{cb} \\
&= 2\omega^{ac} \wedge \omega^{cb} E_{ab}.
\end{aligned}
$$

Comparing this with (8), we see that

(11) $$d\omega = -(1/2)[\omega, \omega].$$

EXERCISE 6. Let ω be any $n \times n$ matrix of 1-forms and let ω^{T} be the transposed matrix. Show that

$$(\omega \wedge \omega)^{\mathrm{T}} = -\omega^{\mathrm{T}} \wedge \omega^{\mathrm{T}}.$$

All the above results also hold for $\mathrm{GL}(n, \mathbb{C})$.

Before computing some more examples, we first discuss a generalized notion of a connection which will be important to us. It differs from the classical definition above of a connection on a principal fiber bundle in that it considers forms that take their values in a larger Lie algebra. We first discuss the situation for a homogeneous space which is a model for the general case. So let G be a Lie group and let H be a closed subgroup. The homogeneous space G/H is the set of left cosets $\{gH\}$ of H in G. Thus G is the total space of a principal fiber bundle with base G/H and fiber group H. Note that H acts on the right in each fiber. The Maurer-Cartan form on G does not provide a connection for this principal fiber bundle since it takes values in $\mathfrak{G}$ instead of in $\mathfrak{H}$. This of course can be easily remedied. We need only choose a complement for $\mathfrak{H}$ in $\mathfrak{G}$ and compose ω with projection into $\mathfrak{H}$ along this complement. However, instead of doing this, we introduce a new type of connection.

DEFINITION. Let P be a principal fiber bundle over M with fiber group H and let G be a Lie group that contains H as a closed subgroup. Let G and P have the same dimension. A 1-form ω with values in $\mathfrak{G}$ is a *Cartan connection* if it satisfies:

$$R_h^*\omega_{uh} = \operatorname{ad}(h^{-1})\omega_u \quad \text{for } h \in H,$$
$$\omega(A^*) = A \quad \text{for } A \in \mathfrak{H},$$
$$\omega(X) \neq 0 \quad \text{for } X \in TP.$$

See [**Ko**] for generalities about Cartan connections and also for several detailed examples.

A Cartan connection on P measures how far M is from being the homogeneous space G/H. To see this, define the curvature Ω of the connection ω to be

$$\Omega = d\omega + \frac{1}{2}[\omega, \omega].$$

If M is the homogeneous space G/H and ω is the Maurer-Cartan connection for G, then $\Omega = 0$. Conversely, if $\Omega = 0$ then M can locally be identified, in an essentially unique way, with a subset of G/H.

We first compute the Maurer-Cartan connection for the group of Euclidean motions and then we compute the connections for two groups and homogeneous spaces which will be important to us later. These are $SU(2, 1)$ which is a subgroup of $GL(3, \mathbb{C})$ and $PGL(2)$ which we take to be a subgroup of $GL(2)$.

The Euclidean motions in the plane can be identified with a subgroup of $GL(3)$ by identifying the motion

$$X \longrightarrow \begin{pmatrix} \cos\theta & -\sin\theta \\ \sin\theta & \cos\theta \end{pmatrix} X + \begin{pmatrix} a \\ b \end{pmatrix}$$

with the matrix

$$\begin{pmatrix} \cos\theta & -\sin\theta & a \\ \sin\theta & \cos\theta & b \\ 0 & 0 & 1 \end{pmatrix}.$$

Note that

$$g_1(g_2 X) = (g_1 g_2)X,$$

where the left-hand side is a composition of motions and the right-hand side is the motion identified with the matrix product of the matrices for g_1 and g_2. For the Maurer-Cartan connection we have

$$\omega = g^{-1}dg = \begin{pmatrix} 0 & -\omega_3 & \omega_1 \\ \omega_3 & 0 & \omega_2 \\ 0 & 0 & 0 \end{pmatrix},$$

where

$$\omega_1 = \cos\theta\, dx + \sin\theta\, dy,$$
$$\omega_2 = -\sin\theta\, dx + \cos\theta\, dy,$$

and

$$\omega_3 = d\theta.$$

From

$$d\omega = -\omega \wedge \omega$$

(or by differentiating directly) we obtain

$$d\omega_1 = \omega_3\omega_2,$$
$$d\omega_2 = -\omega_3\omega_1,$$
$$d\omega_3 = 0.$$

These should be compared to (3) and the equation in Exercise 1 with K set equal to zero.

For our other computations we start with the observation that the Maurer-Cartan connection of a group when restricted to a Lie subgroup gives the Maurer-Cartan connection of the subgroup. Let this subgroup be the group preserving the quadratic form given by some matrix E. That is, let

$$G = \{B: B^*EB = E\}.$$

Then

$$B^{-1} = E^{-1}B^*E$$

and so the Maurer-Cartan connection of G is

$$\omega = B^{-1}\,dB = E^{-1}B^*E\,dB.$$

We want to look at ω more closely. Let $C = E^{\mathrm{T}}$ and let

$$F(Z, W) = Z \cdot C\overline{W} = Z_j C_{jk} \overline{W}_k$$

for vectors Z and W in $\mathbb{C}^n$. For matrices A and B let $F(A, B)$ be the matrix

$$(12) \qquad (F(A, B))_{jk} = F(A_j, B_k) = A_{rj} C_{rs} \overline{B}_{sk}.$$

That is,

$$(13) \qquad F(A, B) = A^{\mathrm{T}} C \overline{B}.$$

Thus

$$(14) \qquad \omega = E^{-1}B^* C^{\mathrm{T}}\,dB = E^{-1}(F(dB, B))^{\mathrm{T}}.$$

To compute the Maurer-Cartan form for $\mathrm{SU}(2, 1)$ we let

$$E = \begin{pmatrix} 0 & 0 & i \\ 0 & 1 & 0 \\ -i & 0 & 0 \end{pmatrix}$$

(see Chapter 2).

We use the notation $B = (B_0, B_1, B_2)$ and

(15) $$E^{-1}(F(dB, B))^{\mathrm{T}} = \begin{pmatrix} \theta_{00} & \theta_{01} & \theta_{02} \\ \theta_{10} & \theta_{11} & \theta_{12} \\ \theta_{20} & \theta_{21} & \theta_{22} \end{pmatrix}.$$

So we have for $j = 0, 1, 2$

(16) $$\begin{cases} \theta_{0j} = iF(dB_j, B_2), \\ \theta_{1j} = F(dB_j, B_1), \\ \theta_{2j} = -iF(dB_j, B_0). \end{cases}$$

These forms are not independent. There are of course a total of eight real independent ones. We now find such a basis. We have from Lemma 2.9 that the elements of $SU(2, 1)$ are completely characterized by

(17) $$\begin{cases} F(B_1, B_1) = 1, \ F(B_0, B_2) = -i, \ F(B_2, B_0) = i, \\ \text{all other } F(B_j, B_k) = 0, \\ \det(B_0, B_1, B_2) = 1. \end{cases}$$

Using our new convention (12), these conditions can be rewritten as

(18) $$F(B, B) = C, \quad \det B = 1.$$

To be consistent with Cartan [Ca2], we introduce a different set of forms, equivalent to $\{\theta_{jk}\}$. Let $\Omega_{00}, \ldots, \Omega_{22}$ be defined by

(19) $$dB_j = \sum_0^2 \Omega_{jk} B_k.$$

From (14) and (18) we see that this implies

(20) $$\Omega = \theta^{\mathrm{T}} = F(dB, B)C,$$

where θ is the matrix on the right-hand side of (15).

Differentiating (18) leads to the equations

$$F(dB_j, B_k) + F(B_j, dB_k) = 0$$

and

$$\det(dB_0, B_1, B_2) + \det(B_0, dB_1, B_2) \\ + \det(B_0, B_1, dB_2) = 0.$$

So we have

$$\Omega_{j\ell} F(B_\ell, B_k) + \overline{\Omega}_{ks} F(B_j, B_s) = 0,$$
$$\Omega_{00} + \Omega_{11} + \Omega_{22} = 0.$$

These lead to

$$\Omega_{02} - \overline{\Omega}_{02} = 0,$$
$$\Omega_{01} - i\overline{\Omega}_{12} = 0,$$
$$\Omega_{00} + \overline{\Omega}_{22} = 0,$$
$$\Omega_{11} + \overline{\Omega}_{11} = 0,$$
$$\Omega_{21} + i\overline{\Omega}_{10} = 0,$$
$$\Omega_{20} - \overline{\Omega}_{20} = 0,$$
$$\Omega_{00} + \Omega_{11} + \Omega_{22} = 0.$$

So Ω_{02} and Ω_{20} are real. As a basis we may take

$$\{\Omega_{02}, \Omega_{20}, \Omega_{00}, \Omega_{01}, \Omega_{21}, \overline{\Omega}_{00}, \overline{\Omega}_{01}, \overline{\Omega}_{21}\}.$$

Following Cartan, we set

(21) $$\Omega = \Omega_{02}, \qquad \Omega_1 = \Omega_{01}, \qquad \Omega_2 = \Omega_{00} - \Omega_{11},$$

$$\Omega_3 = -\Omega_{21}, \qquad \Omega_4 = -\Omega_{20}.$$

So Ω and Ω_4 are real and for our basis for the Maurer-Cartan forms on $SU(2, 1)$ we take

(22) $$\{\Omega, \Omega_1, \Omega_2, \Omega_3, \Omega_4\},$$

where it is understood that for the complex-valued forms $\Omega_1, \Omega_2, \Omega_3$, we take both the real and imaginary parts. So we do end up with eight real 1-forms.

LEMMA 2. *These differential forms satisfy the equations*

$$d\Omega = i\Omega_1\overline{\Omega}_1 - \Omega(\Omega_2 + \overline{\Omega}_2),$$
$$d\Omega_1 = -\Omega_1\Omega_2 - \Omega\Omega_3,$$
$$d\Omega_2 = 2i\Omega_1\overline{\Omega}_3 + i\overline{\Omega}_1\Omega_3 - \Omega\Omega_4,$$
$$d\Omega_3 = -\Omega_1\Omega_4 - \overline{\Omega}_2\Omega_3,$$
$$d\Omega_4 = i\Omega_3\overline{\Omega}_3 + \Omega_4(\Omega_2 + \overline{\Omega}_2).$$

PROOF. As we shall see in a minute, these are easily verified directly. But instead we want to use the structure equations. Since θ is the Maurer-Cartan form, we know it satisfies

$$d\theta = -\theta \wedge \theta.$$

Since $\Omega = \theta^T$ we have (recalling exercise 6)

$$d\Omega = \Omega \wedge \Omega.$$

When the matrix Ω is expressed in terms of the basis (22), we obtain

$$
(23) \qquad \Omega = \begin{pmatrix} \frac{1}{3}(\overline{\Omega}_2 + 2\Omega_2) & \Omega_1 & \Omega \\ i\overline{\Omega}_3 & \frac{1}{3}(\overline{\Omega}_2 - \Omega_2) & i\overline{\Omega}_1 \\ -\Omega_4 & -\Omega_3 & \frac{1}{3}(-2\overline{\Omega}_2 - \Omega_2) \end{pmatrix}.
$$

The lemma follows from computing various components of $\Omega \wedge \Omega$.
To see how the direct verification would go we compute $d\Omega$:

$$
d\Omega = d\Omega_{02} = d\theta_{20} = -\operatorname{id}(F(dB_0, B_0)) = iF(dB_0, dB_0).
$$

Since $dB_0 = \Omega_{0j}B_j$, we have

$$
\begin{aligned}
d\Omega &= i\Omega_{0j} \wedge \overline{\Omega}_{ok} F(B_j, B_k) \\
&= i\{\Omega_{00} \wedge \overline{\Omega}_{02}(-i) + \Omega_{01} \wedge \overline{\Omega}_{01} + \Omega_{02} \wedge \overline{\Omega}_{00}(i)\} \\
&= i\Omega_1 \wedge \overline{\Omega}_1 - \Omega(\Omega_2 + \overline{\Omega}_2).
\end{aligned}
$$

The Maurer-Cartan form on $SU(2, 1)$ becomes the Cartan connection for any homogeneous space of this group. We have seen in Chapter 2 that $SU(2, 1)$ acts transitively on the hyperquadric Q and has the subgroup H as the isotropy group of the origin. Thus Q can be given as the homogeneous space

$$
Q = SU(2, 1)/H.
$$

We shall soon see that the corresponding Cartan connection gives the CR structure on Q. Then in the next chapter, we will give quite a different construction of this connection. The basis of Cartan's study of strongly pseudoconvex three-dimensional CR manifolds is the extension of that construction to these more general manifolds. From such an extension, all the CR properties of the underlying manifold can be read off, at least in theory.

We may use any lift $Q \to SU(2, 1)$, or indeed any map, to pull back the forms $\Omega, \Omega_1, \Omega_2, \Omega_3, \Omega_4$ to Q. Recall that in Chapter 2 we showed that any such lift is given by functions $A_j(x) \in \mathbb{C}^3$, for $j = 0, 1, 2$, and for $x \in Q$, which satisfy (thinking of Q as a subset of $\mathbb{CP}^2$ and using homogeneous coordinates)

$$
(24) \qquad \begin{cases} [A_0(x)] = x, \\ [A_2(x)] \in Q, \ [A_2(x)] \neq [A_0(x)], \\ [A_1(x)] \text{ is the polar of the line determined by } [A_0(x)] \text{ and } [A_2(x)], \\ F(A_1, A_1) = 1, \ F(A_2, A_0) = i, \ F(A_0, A_2) = -i. \end{cases}
$$

We have from (15), (20), and (21) that

(25)
$$\begin{cases} \Omega = -iF(dA_0, A_0), \\ \Omega_1 = F(dA_0, A_1), \\ \Omega_2 = iF(dA_0, A_2) - F(dA_1, A_1), \\ \Omega_3 = -F(dA_2, A_1), \\ \Omega_4 = -iF(dA_2, A_2). \end{cases}$$

We digress for a moment to consider a special lift which gives particularly simple pull-backs. Recall that at the end of Chapter 2 we saw that

(26)
$$\begin{pmatrix} 1 & 0 & 0 \\ z & 1 & 0 \\ u + \frac{i}{2}|z|^2 & i\bar{\zeta} & 1 \end{pmatrix}$$

is an element of $SU(2, 1)$ which projects to $(z, u) \in Q$. The pull-backs of Ω_2, Ω_3, and Ω_4 are all zero while Ω and Ω_1 pull back to

$$\omega = du + \frac{i}{2}(z d\bar{z} - \bar{z} dz) \quad \text{and} \quad \omega_1 = dz.$$

Thus we have the following result, which indicates that the homogeneous structure of Q will be important to us. The remarks after the next lemma show that this result holds for any lift and not just this special one.

LEMMA 3. $\{\omega, \omega_1\}$ gives the CR structure on Q.

That ω is the first of the CR forms can also be seen geometrically. We have

$$dA_0(x) = \omega_{00}A_0(x) + \omega_1 A_1(x) + \omega A_2(x),$$

for any choice of frame. For an infinitesimal variation of A_0 in the complex 2-plane determined by A_0 and A_1, we have $\omega = 0$. For such a variation in A_0, $[A_0]$ varies in the complex line determined by $[A_0]$ and $[A_1]$. Since $[A_1]$ is a polar point, this complex line is tangent to Q at $[A_0]$. That is, this complex line is the real 2-plane H. So $\omega = 0$ does give the characteristic 2-plane of the CR structure.

Now let $(A_0, A_1. A_2)$ be any lift of Q to $SU(2, 1)$ and let $\omega, \omega_1, \ldots$ be the pull-back of $\Omega, \Omega_1, \ldots$. We know from Lemma 2.15 that local coordinates for $SU(2, 1)$ are given by

(27)
$$\begin{cases} B_0 = \tau A_0(x), \\ B_1 = \frac{\tau}{\lambda}(A_1(x) + i\bar{\mu}A_0(x)), \\ B_2 = \frac{\tau}{|\lambda|^2}(A_2(x) - \mu A_1(x) - (\rho + \frac{1}{2}i|\mu|^2)A_0(x)), \end{cases}$$

where $\tau^3 = \lambda^2\bar{\lambda}$. That is, B also provides a lift of Q to $SU(2, 1)$. We now compute the pull-back to Q of the forms $\{\Omega, \Omega_1, \Omega_2, \Omega_3, \Omega_4\}$.

We have

$$\begin{aligned}
\Omega &= -iF(dB_0, B_0) \\
&= -iF(A_0 d\tau + \tau dA_0, \tau A_0) \\
&= -i|\tau|^2 F(dA_0, A_0) \\
&= -i|\tau|^2 \omega_{02} F(A_2, A_0) \\
&= |\tau|^2 \omega \\
&= |\lambda|^2 \omega.
\end{aligned}$$

Also,

$$\begin{aligned}
\Omega_1 &= F(dB_0, B_1) \\
&= \overline{\left(\frac{\tau}{\lambda}\right)} F(A_0 d\tau + \tau dA_0, A_1 + i\overline{\mu}A_0) \\
&= \overline{\left(\frac{\tau}{\lambda}\right)} \tau \{\omega_{01} F(A_1, A_1) + \omega_{02}(-i\mu)F(A_2, A_0)\} \\
&= \frac{|\tau|^2}{\overline{\lambda}}(\omega_1 + \mu\omega) \\
&= \lambda(\omega_1 + \mu\omega).
\end{aligned}$$

The others can be done in the same way. Here is the complete result.

LEMMA 4. *A basis for the Maurer-Cartan form on* $SU(2, 1)$ *is given by*

$$\Omega = |\lambda|^2 \omega,$$

$$\Omega_1 = \lambda(\omega_1 + \mu\omega),$$

$$\Omega_2 = \frac{d\lambda}{\lambda} + \omega_2 - i\mu\overline{\omega}_1 - 2i\overline{\mu}\omega_1 + \left(\rho - \frac{3}{2}i|\mu|^2\right)\omega,$$

$$\Omega_3 = \frac{1}{\lambda}\left(d\mu + \omega_3 + \mu\overline{\omega}_2 + i\mu^2\overline{\omega}_1 + (\rho + \frac{1}{2}i|\mu|^2)\omega_1 \right.$$
$$\left. + (\rho + \frac{1}{2}i|\mu|^2)\mu\omega\right),$$

$$\Omega_4 = \frac{1}{|\lambda|^2}\left(d\rho + \frac{i}{2}(\mu d\overline{\mu} - \overline{\mu}d\mu) + \omega_4 + i\mu\overline{\omega}_3 - i\overline{\mu}\omega_3 \right.$$
$$+ (\rho + \frac{1}{2}i|\mu|^2)\omega_2 + (\rho - \frac{1}{2}i|\mu|^2)\overline{\omega}_2$$
$$- i\overline{\mu}(\rho + \frac{i}{2}|\mu|^2)\omega_1 + i\mu(\rho - \frac{i}{2}|\mu|^2)\overline{\omega}_1$$
$$\left. + (\rho^2 + \frac{1}{4}|\mu|^4)\omega\right).$$

EXERCISE 7. Prove one of the last three formulas.

REMARKS. (1) These same equations, in a more general context, contain all the geometric information about a CR structure. We explain this in the next chapter. Also see p. 1296 of [Ca2, I].

(2) Note that $\{\omega, \omega_1\}$ and $\{|\lambda|^2\omega, \lambda(\omega_1 + \mu\omega)\}$ define the same CR structure. It follows that even for our general mapping $A(x)$ we still have

that $\{\omega, \omega_1\}$ defines the usual CR structure on Q. Thus Lemma 3, which was proved only for the mapping (25), holds for all lifts.

We can look at this lemma from a different viewpoint. We start with any 1-forms ω and ω_1 that define the CR structure on Q. We introduce variables λ, μ, and ρ and obtain an eight-dimensional fiber bundle B with base Q. It is thus possible to find 1-forms ω_2, ω_3, and ω_4 such that $\Omega, \ldots, \Omega_4$ defined as above give a Cartan connection (remember to take real and imaginary parts, so as to end up with eight 1-forms) that satisfies

$$d\Omega = \Omega \wedge \Omega.$$

Hence B is isomorphic to $SU(2, 1)$ and the induced map $Q \to Q$ is CR.

§3. **Projective structures.** Our next example of the Maurer-Cartan connection on a group and the corresponding Cartan connection on a homogeneous space is for the projectivized linear group $PGL(n)$. This will lead us to a discussion of projective structures on curves. We have already seen, in the previous chapter, that the Moser normal form provides us with a projective parametrization on each chain. In Chapter 8, we show how Cartan's approach also produces chains and projective structures, and then compare these two sets of objects.

DEFINITION. The *projective linear group* $PGL(n)$ is the quotient of the general linear group $GL(n)$ by the normal subgroup of multiples of the identity.

We write

$$PGL(n) = GL(n)/\{\lambda I; \ \lambda \neq 0\}.$$

We always think of $GL(n)$ as the group of nonsingular $n \times n$ matrices. Thus each element in $PGL(n)$ is a family of matrices $\{\lambda M : M \in G(n), \ \lambda \in \mathbb{R} - \{0\}\}$. We obtain a realization of $PGL(n)$ by making a convenient choice of λ. We will be interested in $PGL(2)$ and we use the following representation:

$$PGL(2) = \{M \in GL(2), \ \det M = \pm 1\}.$$

Thus $PGL(2)$ consists of two components. For convenience we only consider the identity component and by abuse of notation we also call this component $PGL(2)$.

So

$$PGL(2) = \left\{ \begin{pmatrix} A & B \\ C & D \end{pmatrix} : AD - BC = 1 \right\}.$$

REMARK. This group is better known as $SL(2, R)$. But since we are interested in its connection with projective structures we shall use our less conventional name.

If we consider $\mathbb{P}^1$ as the set of all lines through the origin in $\mathbb{R}^2$, parametrized by the homogeneous coordinates $[\alpha, \beta] = \{(x, y) \mid (x, y) = \lambda(\alpha, \beta)\}$, and $\mathbb{R}^1 \subset \mathbb{P}^1$ given by $\{[\alpha, 1], \ \alpha \in \mathbb{R}^1\}$, then $g = \begin{pmatrix} A & B \\ C & D \end{pmatrix} \in PGL(2)$ acts on $\mathbb{P}^1$ by $g[\alpha, \beta] = [A\alpha + B\beta, C\alpha + D\beta]$ and acts on $\mathbb{R}^1$ by $g(\alpha) =$

$(A\alpha + B)/(C\alpha + D)$. The condition $AD - BC = +1$ means that we are considering only the orientation preserving projective transformations of $\mathbb{R}^1$. As is usual with matrix multiplication, we have that this action on $\mathbb{R}^1$ satisfies $gh(\alpha) = g \circ h(\alpha)$ where the left-hand side is group multiplication and the right-hand side is the composition of functions.

The Lie algebra of $PGL(2)$ is given by its tangent space at the identity as a submanifold of $GL(2)$. Note that

$$\begin{pmatrix} 1 & t \\ 0 & 1 \end{pmatrix}, \qquad \begin{pmatrix} 1 & 0 \\ t & 1 \end{pmatrix}, \qquad \begin{pmatrix} 1+t & 0 \\ 0 & \frac{1}{1+t} \end{pmatrix}$$

give three independent curves in $PGL(2)$ which pass through the identity. Thus the Lie algebra of $PGL(2)$ is generated by

$$\begin{pmatrix} 0 & 1 \\ 0 & 0 \end{pmatrix}, \qquad \begin{pmatrix} 0 & 0 \\ 1 & 0 \end{pmatrix}, \qquad \begin{pmatrix} 1 & 0 \\ 0 & -1. \end{pmatrix}$$

We can easily compute the Maurer-Cartan connection:

$$\omega_{MC} = g^{-1}dg = \begin{pmatrix} D & -B \\ -C & A \end{pmatrix} \begin{pmatrix} dA & dB \\ dC & dD \end{pmatrix} = (\Lambda_{ij})$$

with

$$\Lambda_{11} = DdA - BdC,$$
$$\Lambda_{12} = DdB - BdD,$$
$$\Lambda_{21} = AdC - CdA,$$
$$\Lambda_{22} = AdD - CdB.$$

We know that only three of these forms can be linearly independent; indeed, $AD - BC = 1$ gives that $\Lambda_{11} + \Lambda_{22} = 0$. To find the structural equation of our group we may compute $\Lambda_{ij}\Lambda_{k\ell}$ and express $d\omega_{MC}$ as a combination of such exterior products. (Or we may substitute directly into (8).)

EXERCISE 8. Compute the three nontrivial products $\Lambda_{ij}\Lambda_{k\ell}$. (Hint: Use $\Lambda_{11} = -\Lambda_{22}$ and $\Lambda_{11}\Lambda_{22} = 0$ to simplify these products.)

The structural equation that results is

$$d\Lambda_{11} = -\Lambda_{12}\Lambda_{21},$$
$$d\Lambda_{12} = -2\Lambda_{11}\Lambda_{12},$$
$$d\Lambda_{21} = 2\Lambda_{11}\Lambda_{21}.$$

Consider next the action of $PGL(2)$ on $\mathbb{R}^1$,

$$g(x) = \frac{Ax + B}{Cx + D}.$$

The isotropy subgroup of the origin is given by

$$H = \left\{ \begin{pmatrix} A & 0 \\ C & D \end{pmatrix} : AD = 1 \right\}.$$

A basis for the Lie algebra of $PGL(2)$ was given above; a basis for the Lie algebra of H is

$$\left\{ \begin{pmatrix} 0 & 0 \\ 1 & 0 \end{pmatrix}, \begin{pmatrix} 1 & 0 \\ 0 & -1 \end{pmatrix} \right\}.$$

Note

$$\omega_{MC} = \Lambda_{11} \begin{pmatrix} 1 & 0 \\ 0 & -1 \end{pmatrix} + \Lambda_{12} \begin{pmatrix} 0 & 1 \\ 0 & 0 \end{pmatrix} + \Lambda_{21} \begin{pmatrix} 0 & 0 \\ 1 & 0 \end{pmatrix}.$$

The homogeneous space of left cosets

$$PGL(2)/H = \{gH, \, g \in PGL(2)\}$$

can be identified with $\mathbb{P}^1$. For

$$g = \begin{pmatrix} A & B \\ C & D \end{pmatrix},$$

each element in gH maps the origin to the point B/D in $\mathbb{R}^1$ (provided D is not zero). So we use homogeneous coordinates to define a map of $PGL(2)/H$ to $\mathbb{P}^1$ by

$$\left\{ \begin{pmatrix} A & B \\ C & D \end{pmatrix} H \right\} \xrightarrow{\pi} [B, D].$$

Further, if $\pi(g_1 H) = \pi(g_2 H)$ then $g_1(0) = g_2(0)$, and so $g_1 = g_2 h$. That is, $\pi(g_1 H) = \pi(g_2 H)$ implies $g_1 H = g_2 H$ and π is a bijection.

EXERCISE 9. Show that the action of $g \in PGL(2)$ on $\mathbb{P}^1$ is the same under this identification as left multiplication by g acting on $PGL(2)/H$.

Consider the principal fiber bundle

$$H - PGL(2)$$
$$\downarrow \pi$$
$$\mathbb{P}^1$$

defined by the identification in the exercise. By using ω_{MC} it is easy to recognize when a submanifold of $PGL(2)$ is a fiber $\pi^{-1}(p)$.

EXERCISE 10. (1) Show that when p is given in homogeneous coordinates by $[B, D]$ then

$$\pi^{-1}(p) = \{g \in PGL(2): g(0) = p\} = \left\{ \begin{pmatrix} A & \lambda B \\ C & \lambda D \end{pmatrix} : \lambda(AD - BC) = 1 \right\}.$$

(2) Show that if

$$\gamma(t) = \begin{pmatrix} A(t) & B(t) \\ C(t) & D(t) \end{pmatrix}$$

is a curve in $\pi^{-1}(p)$ and $p = [B, D]$ then

$$\frac{d\gamma}{dt} = \dot{A}\frac{\partial}{\partial A} + \dot{\lambda B}\frac{\partial}{\partial B} + \dot{C}\frac{\partial}{\partial C} + \dot{\lambda D}\frac{\partial}{\partial D}$$

is its tangent vector thought of as an element of the tangent space to the 2×2 matrices. Thus

$$\Lambda_{12}\big|_{\pi^{-1}(p)} = 0.$$

(This says that

$$\omega_{MC}\big|_{\pi^{-1}(p)}$$

takes its values in the Lie algebra of H and thus also follows from the definition of a Maurer-Cartan connection.)

Conversely, show that if M^2 is a submanifold of PGL(2) and $\Lambda_{12}|_{M^2} = 0$ then $M^2 = \pi^{-1}(p)$ for some $p \in \mathbb{P}^1$. (Note that $\Lambda_{12} = 0$ does define a family of two-dimensional submanifolds, since $\Lambda_{12} \wedge d\Lambda_{12} = 0$.)

We have now motivated the following definition.

DEFINITION. A three-dimensional manifold B together with three independent one-forms $\Lambda_1, \Lambda_2, \Lambda_3$ is said to be a *projective structure* if

$$d\Lambda_1 = -\Lambda_2\Lambda_3,$$
$$d\Lambda_2 = -2\Lambda_1\Lambda_2,$$
$$d\Lambda_3 = 2\Lambda_1\Lambda_3.$$

A one-dimensional manifold Γ is said to *admit the projective structure B* if B is a bundle over Γ and $\Lambda_2|_{\pi^{-1}(p)} = 0$ for all $p \in \Gamma$ where $\pi\colon B \to \Gamma$ is the bundle projection.

EXERCISE 11. Show that given a projective structure B, a point $p \in B$, and a point $q \in \text{PGL}(2)$, there is a unique map $\Psi\colon B \to \text{PGL}(2)$ such that $\Psi(p) = q$ and $\Psi^*(\Lambda_{11}) = \Lambda_1$, $\Psi^*(\Lambda_{12}) = \Lambda_2$, and $\Psi^*(\Lambda_{21}) = \Lambda_3$.

There is a more intuitive notion of when a one-dimensional manifold has a projective structure; namely, when it is locally identified with $\mathbb{P}^1$.

DEFINITION. A one-dimensional manifold Γ has a *projective parametrization* if there is given an open covering U_j of Γ and a family of diffeomorphisms $f_j\colon \mathcal{O}_j \to U_j$, with $\mathcal{O}_j$ an open subset of $\mathbb{R}^1$ (or, equivalently, of $\mathbb{P}^1$) such that any two diffeomorphisms with intersecting ranges differ by a projective transformation.

To differ by a projective transformation means, of course, that given f_1 and f_2 there exist constants a, b, c, and d such that

$$f_2\left(\frac{ax+b}{cx+d}\right) = f_1(x)$$

for all x with $f_1(x) \in U_2$.

Cartan's approach will give us, in a natural way, a projective structure on chains. Proposition 4.5 showed that Moser's approach produced a projective parametrization. We will see in Chapter 8 that these coincide. But first we need to discuss the equivalence of the above two definitions. So let us start with a projective parametrization of Γ and try to find a bundle over Γ with the appropriate one-forms. It will be convenient to realize this bundle as a jet bundle. In fact, this will be good practice for us because in the next chapter another, and more complicated, jet bundle will be used to realize Cartan's construction. For now we define only the jet bundle for a one-dimensional manifold and defer the general definition.

So let N and M be one-dimensional with coordinates t and s where t varies over a neighborhood of zero. A map $f\colon N \to M$ is represented in

these coordinates by a function $s = f(t)$. Let $j_n(f) = (f(0), f'(0), \ldots,$
$f^{(n)}(0))$ and let $f \sim_{(n)} g$ mean that $j_n(f) = j_n(g)$. Denote by $[f]_{(n)}$ the
set $\{g: N \to M \mid g \sim_{(n)} f\}$ and by $J^n(N, M)$ the set $\{[f]_{(n)} \mid f: N \to M,$
f a diffeomorphism near $0\}$. (There are other jet bundles in common use.
For example, we may drop the restriction that f is a diffeomorphism or we
may take jets at all points rather than at the origin.)

EXERCISE 12. Show that the equivalence relation is independent of the
choice of coordinates as long as the origin is fixed in N and also show that
$J^n(N, M)$ is a fiber bundle over M of total dimension $n + 1$.

In fact, a local coordinate chart on, say, $J^2(N, M)$ can be given by

$$(x, u, v) \to [x + ut + \frac{1}{2}vt^2].$$

This means that the point $(x, u, v) \in \mathbb{R}^3$ maps to the equivalence class at
the origin of the map $s = f(t) = x + ut + \frac{1}{2}vt^2$. Of course, we must have
$u \neq 0$ since f is always restricted to be a diffeomorphism.

Now take N to be $\mathbb{R}^1$ with t being the usual coordinate. With this
understood, the bundle $J_0^n(\mathbb{R}, M)$ has canonical one-forms $\theta_1, \ldots, \theta_n$. We
want to determine these forms explicitly when $n = 2$. See [Ko] for a general
treatment. We shall see that there are many choices of a one-form Λ such
that θ_1, θ_2, and Λ make $J^2(\mathbb{R}, M)$ a projective structure over N. A
projective parametrization of M permits a unique choice of λ.

Let p be a point of $J^n(\mathbb{R}, M)$ and choose any function $f: \mathbb{R} \to M$ such
that $[f]_{(n)} = p$. Note that f induces by composition a map $J^k(\mathbb{R}, \mathbb{R}) \to$
$J^k(\mathbb{R}, M)$ for each $k \leq n$. This map depends on the particular choice of f,
but as we now show the induced map $T_e J^k(\mathbb{R}, \mathbb{R}) \to T_p J^k(\mathbb{R}, M)$ depends
only on p for each $k \leq n - 1$. Here e denotes $[t]_{(k)}$, the "identity" map.
We are interested only in $n = 2$, so we limit ourselves to this case. We take
as coordinates

$$(s_0, s_1, s_2) \longrightarrow [s_0 + s_1 t + \frac{1}{2}s_2 t^2] \in J^2(\mathbb{R}, \mathbb{R}),$$

$$(x, u, v) \longrightarrow [x + ut + \frac{1}{2}vt^2] \in J^2(\mathbb{R}, M).$$

Let $p = [\overline{x} + \overline{u}t + \frac{1}{2}\overline{v}t^2]$ and choose some $f(t) = \overline{x} + \overline{u}t + \frac{1}{2}\overline{v}t^2 + t^3 g(t)$. Note
$\overline{u} \neq 0$. The composition map $J^2(\mathbb{R}, \mathbb{R}) \to J^2(\mathbb{R}, M)$ is defined by

$$(28) \quad \begin{cases} (s_0, s_1, s_2) = [s_0 + s_1 t + \frac{1}{2}s_2 t^2]_2 \\ \qquad \to [f(s_0 + s_1 t + \frac{1}{2}s_2 t^2)]_2 \\ \qquad = (x(s), u(s), v(s)). \end{cases}$$

Explicitly,

$$x = f(s_0),$$
$$u = f'(s_0)s_1,$$
$$v = f''(s_0)s_1^2 + f'(s_0)s_2.$$

Clearly the image depends not only on $\overline{x}, \overline{u}, \overline{v}$ but also on $g(t)$. Note that the identity e in $J^2(\mathbb{R}, \mathbb{R})$ is given by $s_0 = 0$, $s_1 = 1$, $s_2 = 0$ and that the differential of (28) at e is given by

$$f_*\left(\frac{2}{\partial s_0}\right) = f'(0)\frac{2}{\partial x} + f''(0)\frac{2}{\partial u} + f'''(0)\frac{2}{\partial v},$$
$$f_*\left(\frac{2}{\partial s_1}\right) = f'(0)\frac{\partial}{\partial u} + 2f''(0)\frac{2}{\partial v},$$
$$f_*\left(\frac{2}{\partial s_2}\right) = f'(0)\frac{\partial}{\partial v}.$$

Now, $f(0) = \overline{x}$, $f'(0) = \overline{u}$, $f''(0) = \overline{v}$, $f'''(0) = 6g(0)$. Hence the differential also depends on g. However, if we consider f instead as giving a map of $J^1(\mathbb{R}, \mathbb{R})$ to $J^1(\mathbb{R}, M)$ then its differential

(29) $$f_*: T_e J^1(\mathbb{R}, \mathbb{R}) \to T_p J^1(\mathbb{R}, M)$$

does not depend on g. Namely, we have

(30) $$[s_0 + s_1 t]_{(1)} \longrightarrow [f(s_0 + s_1 t)]_{(1)}$$

and

(31) $$\begin{cases} f_*\left(\dfrac{2}{\partial s_0}\right) = \overline{u}\dfrac{2}{\partial x} + \overline{v}\dfrac{2}{\partial u}, \\ f_*\left(\dfrac{2}{\partial s_1}\right) = \overline{u}\dfrac{2}{\partial u}. \end{cases}$$

Note that the map

$$x = f(s_0), \quad u = s_1 f'(s_0),$$

is a diffeomorphism near $s_0 = 0$, $s_1 = 1$, and thus (29) is an isomorphism. This of course is also clear from (31). This isomorphism depends on p but not on the choice of f.

We now define the canonical 1-forms.

Let

$$\pi: J^2(\mathbb{R}, M) \to J^1(\mathbb{R}, M),$$
$$\pi(s_0, s_1, s_2) = (s_0, s_1),$$

be the natural projection. Consider some $S \in T_p J^2(\mathbb{R}, M)$. So

$$S = \left([x + \overline{u}t + \frac{1}{2}\overline{v}t^2]_{(2)}, \ \alpha\frac{2}{\partial x} + \beta\frac{2}{\partial u} + \gamma\frac{2}{\partial v}\right).$$

Take as above some f with $[f]_{(2)} = p$. Since (29) is an isomorphism there is a unique V with $f_*V = \pi_*S$. We take the coefficients of V to be our canonical 1-forms:

DEFINITION. The *canonical 1-forms* on $J^2(\mathbb{R}, M)$ are defined by

$$f_* \left(\theta_0(S) \frac{2}{\partial s_0} + \theta_1(S) \frac{2}{\partial s_1} \right) = \pi_* S$$

where $S \in T_p J^2(\mathbb{R}, M)$ and f is any function with $[f] = p$.

It is easy to compute these 1-forms. Using (31), the above equation yields

$$\theta_0(S) \left(u \frac{2}{\partial x} + v \frac{2}{\partial u} \right) + \theta_1(S) u \frac{2}{\partial u} = S_0 \frac{2}{\partial x} + S_1 \frac{2}{\partial u}.$$

So

(32) $\theta_0 = u^{-1} dx$ and $\theta_1 = u^{-1} du - v u^{-2} dx$.

Note that

$$d\theta_0 = \theta_0 \wedge \theta_1.$$

We now find a third form on $J^2(\mathbb{R}, M)$ such that

(33) $d\theta_1 = -2\theta_0 \theta_2,$
                         $d\theta_2 = \theta_1 \theta_2.$

This would show that $J^2(\mathbb{R}, M)$ provides a projective structure on M.

EXERCISE 13. Let θ_0 and θ_1 be given by (32). Then θ_2 satisfies (33) if and only if it is of the form

$$\theta_2 = \left(-\frac{1}{4} \frac{v^2}{u^3} + c(x)u \right) dx + \frac{v}{u^2} du - \frac{1}{2u} dv,$$

where $C(x)$ is an arbitrary function.

If we now set

$$\Lambda_1 = \frac{1}{2} \theta_1,$$
$$\Lambda_2 = \theta_0,$$
$$\Lambda_3 = \theta_2,$$

then

$$d\Lambda_1 = -\Lambda_2 \Lambda_3,$$
$$d\Lambda_2 = -2\Lambda_1 \Lambda_2,$$
$$d\Lambda_3 = 2\Lambda_1 \Lambda_3,$$

and $\Lambda_2|_{\pi^{-1}(p)} = 0$ so M admits this projective structure. This is for any choice of $C(x)$. Now we use the projective parametrization to find a unique choice.

Let $f \colon \mathcal{O} \to U$ be one element of the family of diffeomorphisms of the projective parametrization. Consider the map

$$\psi \colon \mathrm{PGL}(2) \to J_0^2(\mathbb{R}, M)$$

given by

$$(34) \qquad \psi \begin{pmatrix} A & B \\ C & D \end{pmatrix} = j^2 \left(f \left(\frac{At + B}{Ct + D} \right) \right).$$

LEMMA 5. 1. $\psi^*(\theta_0) = \Lambda_{12}$, $\psi^*(\theta_1) = 2\Lambda_{11}$.
2. There is a unique θ_2 for which $\psi^*(\theta_2) = \Lambda_{21}$.
θ_2 remains unchanged if f is replaced by any diffeomorphism projectively equivalent to f.

PROOF. To work with $J^2(\mathbb{R}, M)$, we introduce a coordinate on M. It is convenient to choose that one that makes f look like the identity map. So now

$$\psi \begin{pmatrix} A & B \\ C & D \end{pmatrix} = j^2 \left(\frac{At + B}{Ct + D} \right)$$

$$= \left(\frac{B}{D}, \frac{1}{D^2}, -\frac{2C}{D^3} \right).$$

Thus

$$\psi^*(\theta_0) = \psi^*(u^{-1}dx) = D^2 d(b/D) = \Lambda_{12}$$

and

$$\psi^*(\theta_1) = \psi^*(u^{-1}du - vu^{-2}dx)$$

$$= -2D^{-1}dD + 2CD^{-1}(DdB - BdD)$$

$$= -2D^{-1} + 2CdB + 2D^{-1}(1 - AD)dD$$

$$= 2CdB - 2AdD$$

$$= -2\Lambda_{22}$$

$$= 2\Lambda_{11}.$$

This proves (1). Since ψ is a diffeomorphism, (2) is automatic. In fact, θ_2 is the one given by $C(x) \equiv 0$.

To prove the last part of the lemma, note that if $\tilde{f}$ is projectively equivalent to f then $\tilde{f}(x) = Mx$ where M is an element of PGL(2). So the map $\tilde{\psi}$ defined by (34) with f replaced by $\tilde{f}$ is given by

$$\tilde{\psi} = \psi \circ M.$$

Thus

$$\tilde{\psi}^* = L_M^* \psi^*$$

and $\tilde{\psi}^*(\theta) = \psi^*(\theta)$ whenever $\psi^*(\theta)$ is left-invariant. This of course is the case when $\psi^*(\theta) = \Lambda_{21}$. This completes the proof of the lemma.

We have now shown how a projective parametrization gives rise to a projective structure. For the converse, we start with a bundle B over a one-dimensional manifold M and 1-forms $\Lambda_1, \Lambda_2, \Lambda_3$ on B satisfying the definition of a projective structure. Fix some point $p \in B$ and some element

$q \in \mathrm{PGL}(2)$. By Exercise 11 (which of course is solved using the Frobenius theorem) there is a unique map

$$\psi : \mathrm{PGL}(2) \to B,$$

with

$$\psi^*(\Lambda_1) = \Lambda_{11}, \quad \psi^*(\Lambda_2) = \Lambda_{12}, \quad \text{and} \quad \psi^*(\Lambda_3) = \Lambda_{21},$$

and

$$\psi(q) = p.$$

Let

$$\pi_1 : \mathrm{PGL}(2) \to \mathrm{PGL}(2)/H = \mathbb{P}^1$$

and

$$\pi_2 : B \to M$$

be the projections. The fibers of π_1 are given by $\Lambda_{12} = 0$, the fibers of π_2 by $\Lambda_2 = 0$. Thus ψ is a fiber preserving map. Hence the quotient map $\psi : \mathbb{P}^1 \to M$ is well-defined and any projective parametrization of $\mathbb{P}^1$ gives a parametrization of M.

EXERCISE 14. Instead of p and q, choose some other points $\tilde{p} \in B$ and $\tilde{q} \in \mathrm{PGL}(2)$. Consider the corresponding map $\tilde{\psi} : \mathbb{P}^1 \to M$. Assume that the ranges of ψ and $\tilde{\psi}$ overlap. Show that there is some $k \in \mathrm{PGL}(2)$ for which $\tilde{\psi}(kqH) = \psi(qH)$ on some neighborhood of $\pi_1 p$. Conclude that ψ and $\tilde{\psi}$ give the same projective parametrization of M.

This way of obtaining a projective parametrization seems to necessitate first finding ψ. In fact, this can be avoided. We outline this here and carry out the computations for an important example in Chapter 8.

Let $\Gamma(\eta)$ be a curve in $\mathrm{PGL}(2)$ that is transverse to the fibers and take η as a parametrization for $\mathbb{P}^1$. This parametrization gives rise to a projective parametrization of $\mathbb{P}^1$ (i.e., a family of projectively equivalent parametrizations) by admitting all other parametrizations

$$\eta_k^*(\eta) = \pi_1 k \Gamma(\eta), \qquad k \in \mathrm{PGL}(2).$$

Now consider any one of the maps

$$\psi : \mathrm{PGL}(2) \to B$$

obtained above and let

$$\sigma(\eta) = \psi(\Gamma(\eta)).$$

Then $\pi_2 \sigma$ provides a parametrization of M. By the above exercise, a different choice of ψ would give some other parametrization

$$\tilde{\sigma}(\eta) = \psi(k\Gamma(\eta)),$$

with $\tilde{\sigma}(\eta)$ being projectively equivalent to $\sigma(\eta)$. Note that we have almost complete freedom in choosing the curve $\Gamma(\eta)$. Now let $V = \Gamma_* \left(\frac{\partial}{\partial \eta} \right)$ be the tangent vector along the curve and define the functions

$$a_1(\eta) = \Lambda_{11}(V), \qquad a_2(\eta) = \Lambda_{12}(V), \qquad a_3(\eta) = \Lambda_{21}(V).$$

Then $\sigma(\eta)$ is completely determined by the system of ordinary differential equations

(35)
$$\Lambda_j \left(\sigma_* \left(\frac{2}{\partial \eta} \right) \right) = a_j(\eta),$$

and $\eta \to \pi_2(\sigma(\eta)) \in M$ is the projective parametrization for M. Of course, we need no knowledge of ψ to solve (35).

CHAPTER 6

Cartan's Construction

In this chapter we present several derivations of the basic object of study in the geometry of CR structures. Starting with M^3 we construct a bundle B^n and differential 1-forms $\omega^1, \ldots, \omega^n$ with the property that two CR manifolds M_1 and M_2 are CR equivalent if and only if there exists a map $\phi: B_1 \to B_2$ such that $\phi^*(\omega_2^j) = \omega_1^j$. This replaces a hard question, deciding when two CR-structures are equivalent, by what should be an easier question, viz., deciding when an overdetermined system of equations has a solution. Compare this with the example at the beginning of Chapter 5. In principle, every geometric property of M is derivable from this bundle and the 1-forms. So we call such a bundle, together with the forms, a "geometric bundle". In particular, any invariant constructed from the forms is a CR invariant and provides a necessary condition for M_1 and M_2 to be equivalent. It is easy to see that B must be at least eight-dimensional. For the maps of B to itself that preserve the forms provide the CR automorphisms of M. Since the space of maps is at most dimension n (see Exercise 5.3 (2)) while at least one M has an eight-dimensional automorphism group (which M?), we must have $n \geq 8$.

We already have a model for the construction of B and the 1-forms, namely, the Cartan connection on the homogeneous space $Q = \mathrm{SU}(2, 1)/H$. Indeed, the existence of geometric bundles has only been established for strictly pseudoconvex structures and these, of course, are precisely the ones modelled on Q. As we go through the construction the reader should keep in mind the equations given in Lemma 5.2.

§1. **The basic construction.** We start by letting some forms ω and ω_1 define the CR structure. Since we are considering only strictly pseudoconvex structures we may assume ω and ω_1 are chosen to satisfy $d\omega = i\omega_1\overline{\omega}_1$ (mod ω). (See Lemma 1.14.) With this restriction, any other choice of forms will satisfy $\tilde{\omega} = |\lambda|^2\omega$, $\tilde{\omega}_1 = \lambda(\omega_1 + \mu\omega)$ for complex functions λ and μ with $\lambda \neq 0$ (Lemma 1.16). Thus the space of all choices of forms gives a bundle $\tilde{B}$ over M of fiber dimension four and total dimension seven. Further $\Omega = |\lambda|^2\omega$ and $\Omega_1 = \lambda(\omega_1 + \mu\omega)$ are well-defined forms on $\tilde{B}$. This means that they do not depend on the choice of ω and ω_1. If we could find four more well-defined and independent forms (since Ω_1 counts as two real forms) we would have that $\widetilde{B}$ together with these forms is the desired geometrical bundle and we could start discussing geometry. But a priori we know that these extra forms cannot exist! As we have already pointed out, this is because the automorphism group of the hyperquadric is of dimension eight and so B would have to have dimension not less than eight. In fact, we start with the seven-dimensional bundle $\widetilde{B}$ and look for the additional forms and, in the midst of a long computation, see that these forms cannot be uniquely defined except through the artifice of introducing a new variable. Adjoining this variable to the local coordinates of $\widetilde{B}$ implicitly defines some eight-dimensional bundle B. It is rather surprising that one does not really need to know what this new bundle is. All of Cartan's computations are done in local coordinates and the only essential point is the knowledge that the objects that result from these local computations are in fact well-defined (on some bundle) and so are independent of the choice of the original ω and ω_1. For example, in Chapter 8 we use the forms on B to define a set of curves on M^3. The equations that give these curves can be written down using local coordinates. Because there is some bundle B with certain properties the curves are well defined. What, precisely, the bundle is need not concern us.

First we go through Cartan's computations, not in general, but just for the hyperquadric Q. Watch for that extra variable to pop up! Limiting ourselves to Q means that, in Cartan's notation, $b = 0$ and many terms are eliminated. Because $Q \subset \mathbb{C}^2$ we shall take $\omega_1 = dz$. Then we outline the computation for general M but we still take $\omega_1 = dz$. So we are thinking of M as an embedded hypersurface. At the end, we remark that this latter condition is unnecessary. We also relate some of this to the Lewy operator. After this we redo Cartan's computations but in place of B with its normalization $d\omega = i\omega_1\overline{\omega}_1$ (mod ω) we use the somewhat different bundle that comes from the normalization $d\omega = i\omega_1\overline{\omega}_1$ (without any $\omega \wedge \Phi$ terms). The eight-dimensional bundle, which we will again call B, is here the jet bundle which allows one differentiation in the ω direction. This procedure brings the bundle to the forefront.

Finally, we give a third variation of our construction. This one is due to Chern and is meant to serve as an introduction to the analogous construction for higher-dimensional CR hypersurfaces, as in [CM].

So we start by specializing to the hyperquadric Cartan's general construction. The important thing to note will be that Ω, Ω_1, Ω_2, Ω_3, and Ω_4 are all well-defined global objects on the appropriate bundle. We start by choosing some ω and ω_1 that define the CR structure of Q and that satisfy $d\omega = i\omega_1\overline{\omega}_1$ with $\omega_1 = dz$. Fix such a choice. We restate Lemma 1.16 in this case.

LEMMA 1. *If Ω and Ω_1 also define the standard CR structure on Q and in addition satisfy*

$$d\Omega = i\Omega_1\overline{\Omega}_1 \quad (\text{mod } \Omega),$$

then there are complex-valued functions λ and μ, with $\lambda \neq 0$, for which

$$\Omega = |\lambda|^2\omega,$$
$$\Omega_1 = \lambda(\omega_1 + \mu\omega).$$

The forms Ω and Ω_1 may be thought of as forms on the space of all choices (rather than as choices of forms on Q) and thought of in this fashion they do not depend on the original ω and Ω_1! Let us explain this. Let

$$\mathscr{S} = \{(x, \hat{\omega}|_x, \hat{\omega}_1|_x): x \in Q, \quad \hat{\omega} \text{ and } \hat{\omega}_1 \text{ give the}$$
$$\text{CR structure at } x \text{ and}$$
$$d\hat{\omega} \equiv i\omega_1\overline{\omega}_1 \ (\text{mod } \omega)$$
$$\text{at } x\}.$$

$\mathscr{S}$ appears to depend on the 1-jet of ω at x but this is not actually the case. For if $\hat{\omega} = \rho\omega$ then $d\hat{\omega} = \rho i\omega_1\overline{\omega}_1 \ (\text{mod } \hat{\omega})$ and the derivatives of ρ play no role. $\mathscr{S}$ is a bundle over M with fiber dimension equal to four. If we choose ω and ω_1 to describe the CR structure then $\hat{\omega} = |\lambda|^2\omega$ and $\hat{\omega}_1 = \lambda(\omega_1 + \mu\omega)$ so λ and μ provide coordinates for the fiber. Now on $\mathscr{S}$ we define two forms. Let T be tangent to $\mathscr{S}$ at the point $(x, \hat{\omega}, \hat{\omega}_1)$ and let $\pi: \mathscr{S} \to Q$ be the projection. Define forms Ω and Ω_1 by

$$\Omega(T) = \hat{\omega}(\pi_*T) \quad \text{and} \quad \Omega_1(T) = \hat{\omega}_1(\pi_*T).$$

These are clearly well-defined (i.e., they do not depend on any choices). In terms of the local coordinates chosen above

$$\Omega = |\lambda|^2\omega \quad \text{and} \quad \Omega_1 = \lambda(\omega_1 + \mu\omega).$$

It is easy to see the transition functions for the bundle $\mathscr{S}$. We shall always consider the coordinate system on M as fixed. Then a choice of local coordinates on $\mathscr{S}$ means a choice of $\tilde{\omega}$ and $\tilde{\omega}_1$ such that $d\tilde{\omega} = i\tilde{\omega} \wedge \tilde{\omega}_1$. For then we have the coordinate map $(\tilde{\lambda}, \tilde{\mu}) \to (|\tilde{\lambda}|^2\tilde{\omega}, \tilde{\lambda}(\tilde{\omega}_1 + \tilde{\mu}\tilde{\omega}))$. Now if $\tilde{\omega} = |\alpha|^2\omega$ and $\tilde{\omega}_1 = \alpha(\omega_1 + \beta\omega)$ then $\tilde{\lambda}(\tilde{\omega}_1 + \tilde{\mu}\tilde{\omega}) = \tilde{\lambda}(\alpha(\omega_1 + \beta\omega) + \tilde{\mu}|\alpha|^2\omega) = \tilde{\lambda}\alpha\omega_1 + \tilde{\lambda}(\beta + \tilde{\mu}|\alpha|^2)\omega = \lambda(\omega_1 + \mu\omega)$, so $\lambda = \tilde{\lambda}\alpha$ and $\mu = \alpha^{-1}\beta + \overline{\alpha}\tilde{\mu}$ give the coordinate transition functions.

Now we make two important observations.

(i) Any form which can be obtained from Ω and Ω_1 (for instance by exterior derivation and contraction) is itself a well-defined form on $\mathscr{S}$.

(ii) To compute with Ω and Ω_1 we may make any choice of ω and ω_1 and then work in local coordinates. Although the results of the computation will be expressed in local coordinates they in fact do not depend upon our choice.

We start with

(1) $$\omega = du + \frac{1}{2}iz\,d\bar{z} - \frac{1}{2}i\bar{z}\,dz \quad \text{and} \quad \omega_1 = dz.$$

Note that we are taking

$$Q = \left\{ (z,w) : \operatorname{Im} w = \frac{1}{2}|z|^2 \right\}.$$

Then for $\Omega = |\lambda|^2\omega$ and $\Omega_1 = \lambda(\omega_1 + \mu\omega)$ we have

(2) $$d\Omega = \left(\frac{d\lambda}{\lambda} + \frac{d\bar{\lambda}}{\bar{\lambda}} \right)\Omega + i\Omega_1\overline{\Omega}_1 - i|\lambda|^2\overline{\mu}\omega_1\omega - i|\lambda|^2\mu\omega\overline{\omega}_1$$

and

(3) $$d\Omega_1 = \frac{d\lambda}{\lambda}\Omega_1 + \frac{d\mu}{\bar{\lambda}}\Omega + i\lambda\mu\omega_1\overline{\omega}_1.$$

Next we introduce

(4) $$\Omega_2 = \frac{d\lambda}{\lambda} + A\omega_1 + B\overline{\omega}_1 + C\omega,$$

(5) $$\Omega_3 = \frac{1}{\bar{\lambda}}(d\mu + D\omega_1 + E\overline{\omega}_1 + F\omega),$$

and try to impose conditions that force the undetermined coefficients to take unique values. As we have said, we are guided by the Cartan connection over $Q = \mathrm{SU}(2,1)/H$ which is given by Lemmas 5.2 and 5.4. We have

(6) $$d\Omega = (\Omega_2 + \overline{\Omega}_2)\Omega + i\Omega_1\overline{\Omega}_1 + |\lambda|^2(A + \overline{B} + i\overline{\mu})\omega\omega_1 + |\lambda|^2(B + \overline{A} - i\mu)\omega\overline{\omega}_1$$

and

(7) $$d\Omega_1 = \Omega_2\Omega_1 + \Omega_3\Omega + \lambda(A\mu - C + D)\omega\omega_1 + \lambda(B\mu + E)\omega\overline{\omega}_1 + \lambda(B + i\mu)\omega_1\overline{\omega}_1.$$

All terms in ω, ω_1, and $\overline{\omega}_1$ are made to vanish by requiring

(8) $$A = -2i\overline{\mu}, \quad B = -i\mu, \quad C - D = -2i|\mu|^2, \quad \text{and} \quad E = i\mu^2.$$

Next we compute $d\Omega_2$ and $d\Omega_3$ and try to uniquely determine the remaining coefficients. It seems easiest to start with $d(\Omega_2 - \overline{\Omega}_2)$. We have

$$\begin{aligned}
d(\Omega_2 - \overline{\Omega}_2) = {} & -3i(\lambda\overline{\Omega}_3 - E\omega_1 - \overline{D}\,\overline{\omega}_1 - \overline{F}\omega)\omega_1 \\
& -3i(\bar{\lambda}\Omega_3 - D\omega_1 - E\overline{\omega}_1 - F\omega)\overline{\omega}_1 \\
& + dC\omega - d\overline{C}\omega + i(C - \overline{C})\omega_1\overline{\omega}_1 \\
= {} & -3i\overline{\Omega}_3\Omega_1 - 3i\Omega_3\overline{\Omega}_1 \\
& + 3i\lambda\mu\overline{\Omega}_3\omega + 3i\bar{\lambda}\overline{\mu}\Omega_3\omega \\
& + 3i\overline{F}\omega\omega_1 + 3iF\omega\overline{\omega}_1 \\
& + (3iD - 3i\overline{D} + i(C - \overline{C}))\omega_1\overline{\omega}_1 \\
& + dC\omega - d\overline{C}\omega.
\end{aligned}$$

We expand $\Omega_3\omega$ and $\overline{\Omega}_3\omega$ and group terms to obtain

$$d(\Omega_2 - \overline{\Omega}_2) = 3i\Omega_1\overline{\Omega}_3 + 3i\overline{\Omega}_1\Omega_3$$
$$+ (3i\mu d\overline{\mu}\omega + 3i\overline{\mu}d\mu\omega + dC\omega - d\overline{C}\omega)$$
$$+ 3i(\overline{F} - \mu\overline{E} - \overline{\mu}D)\omega\omega_1$$
$$+ 3i(F - \mu\overline{D} - \overline{\mu}E)\omega\overline{\omega}_1$$
$$+ (3iD - 3i\overline{D} + iC - i\overline{C})\omega_1\overline{\omega}_1.$$

We first look at the $\omega_1\overline{\omega}_1$ term. We already had $D = C + 2i|\mu|^2$; so to make this term zero we take

$$(9) \qquad C = \rho - \frac{3}{2}i|\mu|^2 \quad \text{and} \quad D = \rho + \frac{1}{2}i|\mu|^2,$$

where ρ is real but otherwise undetermined. Now note that

$$dC - d\overline{C} = -3i(\overline{\mu}d\mu + \mu d\overline{\mu}).$$

Thus the term within the first pair of parentheses is also gone! Finally, the $\omega\omega_1$ and $\omega\overline{\omega}_1$ terms become zero when we set

$$(10) \qquad F = \mu\rho + \frac{1}{2}i|\mu|^2\mu.$$

The only unknown coefficient left is ρ. We have

$$(11) \qquad d(\Omega_2 - \overline{\Omega}_2) = 3i\Omega_1\overline{\Omega}_3 + 3i\overline{\Omega}_1\Omega_3.$$

Now we look at $\Omega_2 + \overline{\Omega}_2$:

$$\Omega_2 + \overline{\Omega}_2 = \frac{d\lambda}{\lambda} + \frac{d\overline{\lambda}}{\overline{\lambda}} - i\overline{\mu}\omega_1 + i\mu\overline{\omega}_1 + 2\rho\omega,$$
$$d(\Omega_2 + \overline{\Omega}_2) = i\Omega_3\overline{\Omega}_1 - i\overline{\Omega}_3\Omega_1$$
$$+ (i\lambda\mu\overline{\Omega}_3 - i\overline{\lambda}\overline{\mu}\Omega_3 - i\overline{F}\omega_1 + iF\overline{\omega}_1 + 2d\rho)\omega$$
$$+ (-i(\overline{C} - 2i|\mu|^2) - i(C + 2i|\mu|^2) + 2i\rho)\omega_1\overline{\omega}_1.$$

This last term is zero automatically because of our choice of C. One cannot require that ρ satisfy the over-determined system $2d\rho + i\lambda\mu\overline{\Omega}_3 - \cdots = 0$, so we must introduce ρ as a new variable on par with λ and μ. This is in accordance with our previous observation that any geometric bundle must have dimension at least eight. We introduce also a new real form

$$(12) \qquad \Omega_4 = \frac{1}{|\lambda|^2}\left(d\rho + \frac{1}{2}(i\lambda\mu\overline{\Omega}_3 - i\overline{\lambda}\overline{\mu}\Omega_3 + iF\overline{\omega}_1 - i\overline{F}\omega_1) + G\omega\right).$$

Thus

$$(13) \qquad d(\Omega_2 + \overline{\Omega}_2) = i\Omega_3\overline{\Omega}_1 - i\overline{\Omega}_3\Omega_1 + 2\Omega_4\Omega.$$

Combining this with (11) we obtain

$$(14) \qquad d\Omega_2 = 2i\Omega_1\overline{\Omega}_3 + i\overline{\Omega}_1\Omega_3 - \Omega\Omega_4.$$

Now only G is undetermined. The variables x, λ, μ, ρ are local coordinates for some bundle over Q. (Here x is the coordinate for Q.) We have already described how the coordinates (λ, μ) change under a different choice of ω and ω_1. If we had originally chosen $\tilde{\omega}$ and $\tilde{\omega}_1$ with

$$\tilde{\omega} = |\alpha|^2 \omega \quad \text{and} \quad \tilde{\omega}_1 = \alpha(\omega_1 + \beta\omega),$$

then the change of coordinates

$$x, \lambda, \mu \longrightarrow x, \tilde{\lambda}, \tilde{\mu}$$

would, as we have seen, be given by

(15) $$\tilde{\lambda} = \alpha^{-1}\lambda \quad \text{and} \quad \tilde{\mu} = \overline{\alpha}^{-1}\mu - |\alpha|^{-2}\beta.$$

It is not so easy to see how ρ changes. The form $\Omega_2 + \overline{\Omega}_2$ is well-defined but its expression in local coordinates,

(16) $$\Omega_2 + \overline{\Omega}_2 = \frac{d\lambda}{\lambda} + \frac{d\overline{\lambda}}{\overline{\lambda}} - i\overline{\mu}\omega_1 + i\mu\overline{\omega}_1 + 2\rho\omega,$$

was derived under the assumption that $\omega_1 = dz$. The derivation would have to be modified if a more general coframe $\{\tilde{\omega}, \tilde{\omega}_1\}$ were used and the resulting expression would involve $\tilde{\rho}$. Setting this expression equal to the right-hand side of (16) then would give the transition function for $\tilde{\rho}$.

In any event, we do have a well-defined eight-dimensional bundle B over M with fiber coordinates (λ, μ, ρ) and eight real 1-forms on B given by Ω, Ω_1, Ω_2, Ω_3, and Ω_4. Note that these forms are clearly independent and are uniquely determined except for G (which occurs in Ω_4).

We now look at $d\Omega_3$. We have

(17) $$\Omega_3 = \frac{1}{\overline{\lambda}}\left(d\mu + \left(\rho + \frac{1}{2}i|\mu|^2\right)\omega_1 + i\mu^2\overline{\omega}_1 + \left(\mu\rho + \frac{1}{2}i|\mu|^2\mu\right)\omega\right).$$

So

$$d\Omega_3 = -\frac{d\overline{\lambda}}{\overline{\lambda}}\Omega_3 + \frac{1}{\overline{\lambda}}d\rho(\omega_1 + \mu\omega)$$

$$+ \frac{1}{\overline{\lambda}}d\mu\left(\frac{1}{2}i\overline{\mu}\omega_1 + 2i\mu\overline{\omega}_1 + \rho\omega + i|\mu|^2\omega\right)$$

$$+ \frac{1}{\overline{\lambda}}d\overline{\mu}\left(\frac{1}{2}i\mu\omega_1 + \frac{i}{2}\mu^2\omega\right) - \frac{1}{2\overline{\lambda}}|\mu|^2\mu\omega_1\overline{\omega}_1.$$

We now express $\overline{\lambda}^{-1}d\overline{\lambda}$ in terms of $\overline{\Omega}_2$, $\overline{\lambda}^{-1}d\mu$ in terms of Ω_3, and $\overline{\lambda}^{-1}d\rho$ in terms of Ω_4 to obtain

$$d\Omega_3 = -\Omega_1\Omega_4 - \overline{\Omega}_2\Omega_3 + A_1\omega d\mu + A_2\omega d\overline{\mu} + A_3\overline{\omega}_1 d\mu$$

$$+ A_4\omega_1 d\overline{\mu} + A_5\omega\omega_1 + A_6\omega\overline{\omega}_1 + A_7\omega_1\overline{\omega}_1.$$

After a certain amount of computation we see that the previously determined values of A, B, etc., automatically give that each A_j is zero, except for A_5. For this term we have

$$A_5 = \frac{1}{4}|\mu|^4 + \rho^2 - G,$$

and so we take

(18)
$$G = \rho^2 + \frac{1}{4}|\mu|^4.$$

Note this quantity, and hence also Ω_4, is real. We now have

(19)
$$d\Omega_3 = -\Omega_1\Omega_4 - \overline{\Omega}_2\Omega_3.$$

We have now determined eight independent real forms on B. Thus $d\Omega_4$ must be expressible in terms of these forms. Instead of a direct computation we proceed as follows. If we take the exterior derivatives of our equations for $d\Omega_2$ and $d\Omega_3$, i.e., equations (14) and (19), we see that

$$d\Omega_4 = i\Omega_3\overline{\Omega}_3 - \Omega_2\Omega_4 - \overline{\Omega}_2\Omega_4 \quad (\text{mod } \Omega)$$

and that the same equality also holds $\text{mod}\,\Omega_1$. But the fact that Ω_4 is real implies that the modulo terms are both zero and

(20)
$$d\Omega_4 = i\Omega_3\overline{\Omega}_3 - \Omega_2\Omega_4 - \overline{\Omega}_2\Omega_4.$$

The forms $\Omega, \ldots, \Omega_4$ together with the underlying bundle give the geometrical bundle which completely describes the CR geometry of Q. Note that the equations for the exterior derivatives of these forms involve only constants; thus the bundle is (locally) a Lie group. Indeed, these forms are identical with those in Lemma 5.4 and so are the components of the Maurer-Cartan forms on $SU(2, 1)$.

We now turn to constructing the geometric bundle for a general strictly pseudoconvex M^3. Following Cartan, we at first assume M^3 is embedded in $\mathbb{C}^2$ and that the restriction to M of the coordinate function z satisfies $dz \wedge d\bar{z} \neq 0$. We take $\omega_1 = dz$ and normalize ω so that

(21)
$$d\omega = i\omega_1\overline{\omega}_1 + b\omega\omega_1 + \bar{b}\omega\overline{\omega}_1.$$

We will be using the following notation:

(22)
$$df = f_0\omega + f_1\omega_1 + f_{\bar{1}}\overline{\omega}_1.$$

EXERCISE 1. Let L be the Lewy operator (so $\omega(L) = 0$, $\omega_1(L) = 0$) normalized by $\overline{\omega}_1(L) = 1$. Show that $f_1 = \overline{L}f$ and $f_{\bar{1}} = Lf$.

EXERCISE 2. Show that for any complex-valued function f, and any forms ω and ω_1 with ω real and $\omega \wedge \omega_1 \wedge \overline{\omega}_1 \neq 0$,

$$\overline{f_0} = (\bar{f})_0,$$
$$\overline{f_1} = (\bar{f})_{\bar{1}},$$
$$\overline{f_{\bar{1}}} = (\bar{f})_1.$$

EXERCISE 3. (to be used in Chapter 8). Show that if $\omega = adu + \beta dz + \overline{\beta}d\bar{z}$ and $\omega_1 = dz$ then

$$f_0 = a^{-1}f_u,$$
$$f_1 = f_z - a^{-1}\beta f_u,$$
$$f_{\bar{1}} = f_{\bar{z}} - a^{-1}\overline{\beta}f_u.$$

Again we start with

(23) $$\Omega = |\lambda|^2 \omega \quad \text{and} \quad \Omega_1 = \lambda(\omega_1 + \mu\omega)$$

and introduce

(24) $$\Omega_2 = \frac{d\lambda}{\lambda} + A\omega_1 + B\overline{\omega_1} + C\omega$$

and

(25) $$\Omega_3 = \frac{1}{\lambda}(d\mu + D\omega_1 + E\overline{\omega_1} + F\omega).$$

In order to obtain

(26) $$d\Omega = i\Omega_1\overline{\Omega}_1 + (\Omega_2 + \overline{\Omega}_2)\Omega$$

and

(27) $$d\Omega_1 = \Omega_2\Omega_1 + \Omega_3\Omega,$$

we need to take

(28) $$A = -(b + 2i\overline{\mu}),$$
(29) $$B = -i\mu,$$
(30) $$C - D = -2i|\mu|^2,$$
(31) $$E = -\mu(\overline{b} - i\mu).$$

Note that when $b \equiv 0$, these reduce to (8).

We now compute

(32) $$d(\Omega_2 - \overline{\Omega}_2) - 3i\Omega_1\overline{\Omega}_3 - 3i\overline{\Omega}_1\Omega.$$

As the coefficient of $\omega_1\overline{\omega}_1$ we obtain, after simplifications using (28) and (31),

$$4i(C - \overline{C}) + (b_{\overline{1}} + \overline{b}_1) - 12|\mu|^2.$$

Note that $b_{\overline{1}}$ is real (this can easily be seen by taking the exterior derivative of $d\omega$) and equals $\overline{b}_1$ (Exercise 2). So we set

(33) $$c = b_{\overline{1}}.$$

To make the coefficient of $\omega_1\overline{\omega}_1$ equal to zero we take

(34) $$C = \rho + \frac{1}{4}ic - \frac{3}{2}i|\mu|^2,$$

where ρ is arbitrary. This of course also determines D,

(35) $$D = C + 2i|\mu|^2 = \rho + \frac{1}{4}ic + \frac{1}{2}i|\mu|^2.$$

Using this value for C, we see that the only nonzero terms in

$$d(\Omega_2 - \overline{\Omega}_2) - 3i\Omega_1\overline{\Omega}_3 - 3i\overline{\Omega}_1\Omega_3$$

are the $\omega\omega_1$ and $\omega\overline{\omega}_1$ terms. Setting

(36) $$F = \frac{1}{2}i\mu|\mu|^2 + \mu\rho - \frac{1}{4}ic\mu + \frac{1}{6}(c_{\overline{1}} - \overline{b}c + 2i\overline{b}_0)$$

makes both of these terms zero. So

(37) $$d(\Omega_2 - \overline{\Omega}_2) = 3i\Omega_1\overline{\Omega}_3 + 3i\overline{\Omega}_1\Omega_3.$$

We admit ρ as a new variable. Hence we may consider Ω, Ω_1, Ω_2, and Ω_3 as now being uniquely determined.

LEMMA 2.

$$(d(\Omega_2 + \overline{\Omega}_2) - i\Omega_1\overline{\Omega}_3 + i\overline{\Omega}_1\Omega_3) \wedge \Omega = 0.$$

PROOF. Take the exterior derivative of each side of

$$d\Omega = i\Omega_1\overline{\Omega}_1 + (\Omega_2 + \overline{\Omega}_2)\Omega,$$

and use

$$d\Omega_1 = \Omega_2\Omega_1 + \Omega_3\Omega.$$

Thus

(38) $$d(\Omega_2 + \overline{\Omega}_2) = i\Omega_1\overline{\Omega}_3 - i\overline{\Omega}_1\Omega_3 + 2\Omega_4\Omega,$$

where Ω_4 is uniquely determined except for an arbitrary term $G\omega$. Note that if we restrict G to be real, then Ω_4 is real. In fact,

(39) $$\Omega_4 = \frac{1}{|\lambda|^2}\left\{d\rho + \frac{i}{2}(\mu d\overline{\mu} - \overline{\mu}d\mu)\right.$$

$$+ \left(\frac{1}{2}|\mu|^2\mu - i\overline{\mu}\rho - b\rho - \frac{1}{2}ib|\mu|^2 + \frac{1}{4}c\overline{\mu} + \frac{1}{2}b_0 - \frac{1}{12}i\ell\right)\omega_1$$

$$+ \left(\frac{1}{2}|\mu|^2\overline{\mu} + i\mu\rho - \overline{b}\rho + \frac{1}{2}i\overline{b}|\mu|^2 + \frac{1}{4}c\mu + \frac{1}{2}\overline{b}_0 + \frac{1}{12}i\overline{\ell}\right)\overline{\omega}_1$$

$$\left. + G\omega\right\},$$

where

(40) $$\ell = c_1 - bc - 2ib_0.$$

From (37) and (38) we have

(41) $$d\Omega_2 = 2i\Omega_1\overline{\Omega}_3 + i\overline{\Omega}_1\Omega_3 + \Omega_4\Omega.$$

Let us gather together what we have so far. The forms

(42)
$$\begin{cases} \Omega = |\lambda|^2\omega, \\ \Omega_1 = \lambda(\omega_1 + \mu\omega), \\ \Omega_2 = \dfrac{d\lambda}{\lambda} + A\omega_1 + B\overline{\omega}_1 + C\omega, \\ \Omega_3 = \dfrac{1}{\overline{\lambda}}(d\mu + D\omega_1 + E\overline{\omega}_1 + F\omega), \\ \Omega_4 = \dfrac{1}{|\lambda|^2}\left\{d\rho + \dfrac{i}{2}(\mu d\overline{\mu} - \overline{\mu}d\mu) + H\omega_1 + \overline{H}\overline{\omega}_1 + G\omega\right\}, \end{cases}$$

with only G not yet determined, satisfy the equations

(43)
$$\begin{cases} d\Omega = i\Omega_1\overline{\Omega}_1 - \Omega(\Omega_2 + \overline{\Omega}_2), \\ d\Omega_1 = -\Omega_1\Omega_2 - \Omega\Omega_3, \\ d\Omega_2 = 2i\Omega_1\overline{\Omega}_3 + i\overline{\Omega}_1\Omega_3 + \Omega_4\Omega, \end{cases}$$

and are the only forms, with Ω and Ω_1 given as above, which do satisfy these equations.

LEMMA 3.

(a) $(d\Omega_3 + \Omega_1\Omega_4 + \overline{\Omega}_2\Omega_3) \wedge \Omega = 0$,

(b) $(d\Omega_3 + \Omega_1\Omega_4 + \overline{\Omega}_2\Omega_3) \wedge \Omega_1 \wedge \overline{\Omega}_1 = 0$.

COROLLARY. $d\Omega_3 + \Omega_1\Omega_4 + \overline{\Omega}_2\Omega_3 = f\Omega\Omega_1 + h\Omega\overline{\Omega}_1$.

PROOF. For (a), start by taking the exterior derivative of

$$d\Omega_1 = -\Omega_1\Omega_2 - \Omega\Omega_3,$$

and use

$$d\Omega_2 = 2i\Omega_1\overline{\Omega}_3 + i\overline{\Omega}_1\Omega_3 - \Omega\Omega_4.$$

For (b), a not too long computation shows that $d\Omega_3 + \Omega_1\Omega_4 + \overline{\Omega}_2\Omega_3$ is a combination, with coefficients depending on all variables, of 2-forms on M. In light of (a), this suffices.

The only coefficient not yet determined is G, the coefficient of ω in Ω_4. It is clear from the Corollary that there is a unique choice of G for which f is zero. In fact, this choice is

(44)
$$\begin{aligned} G = &\frac{11}{48}c^2 + \frac{1}{6}\left(|b|^2c + b\overline{\ell} + \overline{b}\ell - g\right) \\ &+ \frac{i}{6}(\mu\ell - \overline{\mu}\overline{\ell}) - \frac{1}{4}c|\mu|^2 + \rho^2 + \frac{1}{4}|\mu|^4, \end{aligned}$$

where

(45)
$$g = c_{1\overline{1}} - \frac{1}{2}ic_0.$$

Note that G is real. We now have

(46)
$$d\Omega_3 = -\Omega_1\Omega_4 - \overline{\Omega}_2\Omega_3 - R\Omega\overline{\Omega}_1.$$

A straightforward computation gives

(47)
$$R = \frac{r}{\lambda\overline{\lambda}^3},$$

where

(48)
$$r = \frac{1}{6}(\overline{\ell}_{\overline{1}} - 2\overline{b}\ell).$$

We now have eight independent real forms in eight variables. Thus $d\Omega_4$ must be expressible in terms of $\Omega, \ldots, \Omega_4$. If we take $d^2\Omega_2$ we derive that

$$d\Omega_4 = i\Omega_3\overline{\Omega}_3 + \Omega_4(\Omega_2 + \overline{\Omega}_2) + \Phi\Omega.$$

To learn something about the 1-form Φ, we substitute this into the expression for $d^2\Omega_3$. The result is that

$$d\Omega_4 = i\Omega_3\overline{\Omega}_3 - (\Omega_2 + \overline{\Omega}_2)\Omega_4 - S\Omega\Omega_1 - \overline{S}\Omega\overline{\Omega}_1$$

where S is some function of $u, z, \lambda, \overline{\lambda}, \mu, \overline{\mu}, \rho$. It is natural to consider R and S as the components of the "curvature" of the CR structure since they measure how far the structure is from the hyperquadric Q.

EXERCISE 4. Show that if $R = 0$ over some open set of M, then S is also zero over this open set. (Hint: Compute $d^2\Omega_3$.)

EXERCISE 5. Show that if $R = 0$ in a neighborhood of some point of M then there is a possibly smaller neighborhood of that point which is CR diffeomorphic to some open subset of Q.

To summarize Cartan's construction:

Let $\{\omega, \omega_1\}$ give the CR structure with $\omega_1 = dz$ and ω normalized by

(49) $$d\omega = i\omega_1\overline{\omega}_1 + b\omega\omega_1 + \overline{b}\omega\overline{\omega}_1.$$

Introduce auxiliary variables λ, μ, ρ with ρ real. Define the 1-forms

(50) $$\begin{cases} \Omega = |\lambda|^2\omega, \\ \Omega_1 = \lambda(\omega_1 + \mu\omega), \\ \Omega_2 = \dfrac{d\lambda}{\lambda} + A\omega_1 + B\overline{\omega}_1 + C\omega, \\ \Omega_3 = \dfrac{1}{\overline{\lambda}}(d\mu + D\omega_1 + E\overline{\omega}_1 + F\omega), \\ \Omega_4 = \dfrac{1}{|\lambda|^2}\left(d\rho + \dfrac{i}{2}(\mu d\overline{\mu} - \overline{\mu}d\mu) + H\omega_1 + \overline{H}\overline{\omega}_1 + G\omega\right), \end{cases}$$

$A, \ldots, H$ given by (28), (29), (34), (35), (31), (36), (44), and (39), respectively. These forms satisfy the equations

(51) $$\begin{cases} d\Omega = i\Omega_1\overline{\Omega}_1 - \Omega(\Omega_2 + \overline{\Omega}_2), \\ d\Omega_1 = -\Omega_1\Omega_2 - \Omega\Omega_3, \\ d\Omega_2 = 2i\Omega_1\overline{\Omega}_3 + i\overline{\Omega}_1\Omega_3 - \Omega\Omega_4, \\ d\Omega_3 = -\Omega_1\Omega_4 - \overline{\Omega}_2\Omega_3 - R\Omega\overline{\Omega}_1, \\ d\Omega_4 = i\Omega_3\overline{\Omega}_3 - (\Omega_2 + \overline{\Omega}_2)\Omega_4 - S\Omega\Omega_1 - \overline{S}\Omega\overline{\Omega}_1, \end{cases}$$

where R is given by (47) and (48) and S has the property in Exercise 4. Further, for $\omega_1 = dz$ and ω any 1-form which, together with ω_1, gives the CR structure, and which is normalized by (49), the only forms that satisfy (51) and the first two equations of (50) are given by the remainder of (50).

The geometric context of this construction is fundamental and beautiful but certainly not immediate. We explore some consequences in the next three chapters. But now, after a few concluding remarks, we want to give two other approaches to this same construction.

First, note that c depends on the first derivatives of b, ℓ on the second, and r on the third.

EXERCISE 6. Let $f(z, w, \overline{z}, \overline{w})$ be the defining function of some strictly pseudoconvex M^3 in $\mathbb{C}^2$. Show that b depends on the third derivatives of f. So the "curvature" r depends on the sixth derivatives of f. (Hint: Start with $\omega = i\partial f$ and then replace ω by $g\omega$ with g chosen so that $d\omega \equiv idz \wedge d\overline{z}$ (mod ω). See Chapter 8, if necessary.) In Lemma 8.7, r is shown to be a multiple of the coefficient of $z^2\overline{z}^4$ in the Moser normal form.

Second, note that Cartan assumed that ω_1 could be chosen to be dz, i.e., that there exists some CR function z on M (and that this function satisfies $dz \wedge d\overline{z} \neq 0$). In fact, Cartan only concerned himself with hypersurfaces M^3 realized in $\mathbb{C}^2$. Certainly, all real analytic hypersurfaces can be so realized (locally); see Chapter 1. But, as we explain in Chapter 10, there exist abstract strictly pseudoconvex CR structures which are not realizable and indeed which admit only the constants as CR functions. Cartan's construction seems not to apply to such structures. But it is possible to redo the construction without assuming $d\omega_1 = 0$. One again gets unique forms $\Omega, \ldots, \Omega_4$ which satisfy the structure equations (50) and (51). However, the explicit values for the coefficients $A, \ldots, H$ are naturally changed. For instance, if we have

$$d\omega_1 = \alpha\omega\omega_1 + \beta\omega\overline{\omega}_1 + \gamma\omega_1\overline{\omega}_1,$$

then

$$A = -(b + 2i\overline{\mu}) + \overline{\gamma}.$$

In fact, each coefficient $A, \ldots, H$ differs from the corresponding expression (28), etc., by a function on M. That is, the values of $A, \ldots$, computed under the assumption that $d\omega_1 = 0$, differ from the values without this assumption only by terms that are independent of λ, μ, and ρ.

Our final remark concerns the Lewy operator. Any complex vector field

$$L = \frac{\partial}{\partial\overline{z}} + B(z, \overline{z}, u)\frac{\partial}{\partial u}$$

defines a CR structure (see Chapter 1).

EXERCISE 7. Show that this CR structure can be given by

$$\omega = k(du - \overline{B}dz - Bd\overline{z}),$$
$$\omega_1 = dz,$$

where

$$k = i(B\overline{B}_u + \overline{B}_{\overline{z}} - \overline{B}B_u - B_z)^{-1},$$

and that

$$d\omega = i\omega_1\overline{\omega}_1 + b\omega\omega_1 + \overline{b}\omega\overline{\omega}_1$$

with

$$b = -\frac{\overline{L}k}{k} - \overline{B}_u.$$

In this way a very sophisticated geometry (see the next three chapters) can be associated to the partial differential operator L.

§2. **Two variations.** Now come the two variations of this construction. They both have the advantage that an underlying eight-dimensional bundle, similar to the one implied in Cartan's construction, is introduced at the beginning. The first treatment uses a certain jet bundle. The second, due to Chern, is somewhat more intrinsic and uses ideas from the theory of G-structures.

So we start with jet bundles. Let E be a fiber bundle over M and let $J^r(M, E)$ be the bundle of r-jets of sections of M into E rather than just of maps of M into the manifold E. Let $\Lambda^r(M)$ denote the bundle of real r-forms on M; a point of $\Lambda^r(M)$ is a pair (x, ω) with $x \in M$ and ω in the rth exterior product of $T_x^*(M)$. Finally, let L denote some choice of the Lewy operator. Set

$$E_1 = \{\omega \in \Lambda^1(M): \omega(L) = 0\},$$
$$E_2 = \{\omega_1 \in \mathbb{C} \otimes \Lambda^1(M): \omega_1(L) = 0, \ \omega_1(\overline{L}) \neq 0\},$$

and

$$\widehat{B} = J^1(M, E_1) \times J^0(M, E_2).$$

Note that $\omega \in E_1$ is real, so also $\omega(\overline{L}) = 0$. The following lemma is clear.

LEMMA 4. *Each section* (ω, ω_1) *of* $J^0(M, E_1) \times J^0(M, E_2)$ *defines the given CR structure. In particular,* $\omega \wedge \omega_1 \wedge \overline{\omega}_1 \neq 0$.

Next note that there is a well-defined map $d: J^1(M, E_1) \to \Lambda^2(M)$ given as follows: Let $q \in J^1(M, E_1)$, so $q = j_x^1(\omega(\tilde{x}))$. Then $dq = d\omega(x)$. We shall think of d as a map $\widehat{B} \to \Lambda^2(M)$ by ignoring the second factor in $\widehat{B}$. Let

$$B = \left\{(j^1(\omega), \omega_1) \in \widehat{B} \ | \ d\omega = i\omega_1\overline{\omega}_1\right\}.$$

We shall show that B is a sub-bundle of $\widehat{B}$. To this end, let us introduce coordinates on $\widehat{B}$. So fix some choice of ω^0, ω_1^0 for such sections. Then we have local coordinates for $\widehat{B}$ given by

$$(x, r, a_1, a_2, a_3, \lambda, \mu) \longrightarrow \left(j_x^1\Big((r + a_i(\tilde{x}_i - x_i))\omega^0(\tilde{x})\Big), \ \lambda(\omega_1^0(x) + \mu\omega^0(x))\right),$$

and in terms of these coordinates $d\omega$ at x is given by $r d\omega^0(x) + a_j \, dx_j \wedge \omega^0(x)$.

Now write

$$dx_j = b_j\,\omega^0 + c_j\,\omega_1^0 + \overline{c}_j\,\overline{\omega}_1^0$$

with b_j real and c_j complex.

EXERCISE 8. Use the fact that $dx_1 \wedge dx_2 \wedge dx_3 \neq 0$ to show the equation $c_j y_j = 0$, y_j real, has a solution set of real dimension one.

The set B is defined by the equation

(52) $$r \, d\omega^0 + a_j \, dx_j \wedge \omega^0 = i|\lambda|^2 (\omega_1^0 + \mu\omega^0) \wedge (\overline{\omega}_1^0 + \overline{\mu}\omega^0),$$

which we rewrite as

$$r \, d\omega^0 + a_j (b_j \, \omega^0 + c_j \, \omega_1^0 + \overline{c}_j \, \overline{\omega}_1^0) \wedge \omega^0 = i|\lambda|^2 (\omega_1^0 + \mu\omega_1^0)(\overline{\omega}_1^0 + \overline{\mu}\omega^0).$$

We are assuming that $d\omega^0 = i\omega_1^0\overline{\omega}_1^0$. So it follows that $r = |\lambda|^2$ and $a_j c_j = i|\lambda|^2\overline{\mu}$. By the exercise, $\{a_j\}$ is determined up to one real dimension. Thus B is a bundle and for fiber coordinates we may take λ, μ, and some real parameter. Let us see exactly what this real parameter is. Express $a_j \, dx_j$ as $t\omega^0 + s\omega_1^0 + \overline{s\omega_1^0}$. Then (52) becomes

$$r \, d\omega^0 + (s\omega_1^0 + \overline{s\omega_1^0}) \wedge \omega^0 = i|\lambda|^2 (\omega_1^0 + \mu\omega^0) \wedge (\overline{\omega}_1^0 + \overline{\mu}\omega^0);$$

so $s = i|\lambda|^2\overline{\mu}$ and t is arbitrary. Thus B consists of $(j_x^1(R\omega^0), j_x^0(\omega_1^0))$ where $dR = t\omega^0 + s\omega_1^0 + \overline{s\omega_1^0}$, with t arbitrary but R, s, and ω_1^0 chosen such that $d(R\omega^0) = i\omega_1^0\overline{\omega}_1^0$ at x. In fact, we shall let $\rho = -\frac{1}{2}\frac{t}{|\lambda|^2}$ and take as fiber coordinates for B the quantities ρ, λ, μ.

Since B is a jet bundle there are special sections which we shall refer to as "good". Recall that a section of $J^1(R^1, R^1)$ is given by $(x, j_x^1(f_x))$ where for each x, f_x is a function of some variable, say $\tilde{x}$, and these functions are chosen so that $f(x)$ and $\frac{\partial f_x(\tilde{x})}{\partial x}\big|_{\tilde{x}=x}$ vary smoothly. The simplest, *but not the only*, way to do this is to take $f_x(\tilde{x}) = f(\tilde{x})$ for one function f. We say a section of $J^1(R^1, R^1)$ is "good" if it is generated in this way by a single function. If as coordinates on $J^1(R^1, R^1)$ we have (x, y, y_1) where $j_x^1(f) = (x, f(x), f'(x))$ then the good sections are precisely those sections that are annihilated by the form $\omega = y_1 dx - dy$. (That is, if the section is $\phi(x) = (x, y(x), y'(x))$ then $\phi^*\omega = 0$.)

DEFINITION. A good section of B over some set $U \in M$ is given by $(j^1(\omega), \omega_1)$ where ω and ω_1 are smooth 1-forms on U that satisfy $d\omega = i\omega_1\overline{\omega}_1$.

In terms of our local coordinates we have that $\lambda(x)$ and $\mu(x)$ are determined by $\omega_1 = \lambda(\omega_1^0 + \mu\omega^0)$. Then $\omega = f\omega^0$ with $f(x) = |\lambda(x)|^2$ and $df = -2\rho|\lambda|^2\omega^0 + i|\lambda|^2\overline{\mu}\omega_1^0 - i|\lambda|^2\mu\overline{\omega}_1^0$. Thus $d|\lambda|^2 = -2\rho|\lambda|^2\omega^0 + i|\lambda|^2\overline{\mu}\omega_1^0 - i|\lambda|^2\mu\overline{\omega}_1^0$. This follows from our formulas above for R and s. Thus the form

$$\frac{d\lambda}{\lambda} + \frac{d\overline{\lambda}}{\overline{\lambda}} + 2\rho\omega^0 - i\overline{\mu}\omega_1^0 + i\mu\overline{\omega}_1^0$$

annihilates all good sections. It is easy to see that, conversely, if $\frac{d\lambda}{\lambda} + \frac{d\bar{\lambda}}{\bar{\lambda}} + \Phi$ annihilates all good sections and Φ annihilates the fiber over each $x \in M$ then $\Phi = 2\rho\omega^0 - i\bar{\mu}\omega_1^0 + i\mu\bar{\omega}_1^0$.

We now show that Cartan's computations can be carried out in B. Thus the bundle appears right at the beginning of our work. We do this for the general strictly pseudoconvex structure. Where possible we avoid tedious computations by proving that certain terms must be zero, rather than just carrying out the computations and noting cancellations. Our computations differ somewhat from Cartan's because he worked with forms such that

$$d\omega = i\omega_1\bar{\omega}_1 + b\omega\omega_1 + \bar{b}\omega\bar{\omega}_1,$$
$$d\omega_1 = 0.$$

In our bundle b is zero, so we will be working with forms satisfying

$$d\omega = i\omega_1\bar{\omega}_1,$$
$$d\omega_1 = \alpha_{01}\omega\omega_1 + \alpha_{0\bar{1}}\omega\bar{\omega}_1 + \alpha_{1\bar{1}}\omega_1\bar{\omega}_1.$$

We end up with the same structural equations Cartan obtained. Thus there is some isomorphism between Cartan's bundle and ours. But since it is rather hard to describe exactly with which bundle Cartan worked, we will not try to investigate this further.

The next result has already been discussed for the hyperquadric. The same computations establish it for the bundle B. Recall (ω^0, ω_1^0) is some choice of forms satisfying $d\omega^0 = i\omega_1^0\bar{\omega}_1^0$ and this choice introduces the local coordinates λ, μ, ρ. From now on let us just use (ω, ω_1) to denote some choice.

LEMMA 5. *The 1-forms* $\Omega = |\lambda|^2\omega$ *and* $\Omega_1 = \lambda(\omega_1 + \mu\omega)$ *are well-defined on* B.

The lemma means that if one chooses $\tilde{\omega}$ and $\tilde{\omega}_1$ in place of ω and ω_1 and so has different local coordinates $\tilde{\lambda}$ and $\tilde{\mu}$ (ρ and $\tilde{\rho}$ are irrelevant here) then $|\tilde{\lambda}|^2\tilde{\omega} = |\lambda|^2\omega$ and $\tilde{\lambda}(\tilde{\omega}_1 + \tilde{\mu}\tilde{\omega}) = \lambda(\omega_1 + \mu\omega)$.

LEMMA 6. *There is a unique form* ϕ *satisfying*

(53)
$$d\Omega = i\Omega_1\bar{\Omega}_1 + \phi\Omega,$$

(54)
$$\phi \text{ vanishes on good sections.}$$

PROOF. For the existence part we try

$$\phi = \frac{d\lambda}{\lambda} + \frac{d\bar{\lambda}}{\bar{\lambda}} + a\omega_1 + b\bar{\omega}_1 + c\omega$$

and see that there is such a form of this type, namely

$$\phi = \frac{d\lambda}{\lambda} + \frac{d\bar{\lambda}}{\bar{\lambda}} - i\bar{\mu}\omega_1 + i\mu\bar{\omega}_1 + 2\rho\omega.$$

Could there be any other form $\tilde{\phi}$ that also satisfies (53) and (54)? Well, we would need $(\phi - \tilde{\phi}) \wedge \Omega = 0$, so $\tilde{\phi}$ could differ from ϕ only in its ω term. But then the condition (54) would force these ω terms to agree.

We are using the notation

$$(55) \qquad d\omega_1 = \alpha_{01}\omega\omega_1 + \alpha_{0\bar{1}}\omega\bar{\omega}_1 + \alpha_{1\bar{1}}\omega_1\bar{\omega}_1.$$

LEMMA 7. *The 1-forms ψ and Ω_3, with $\psi = -\overline{\psi}$, satisfy*

$$(56) \qquad d\Omega_1 = \frac{1}{2}(\phi + \psi)\Omega_1 + \Omega_3\Omega$$

if and only if

$$(57) \qquad \psi = \frac{d\lambda}{\lambda} - \frac{d\bar{\lambda}}{\bar{\lambda}} + A\omega_1 - \overline{A\omega_1} + C\omega$$

and

$$(58) \qquad \Omega_3 = \frac{1}{\bar{\lambda}}(d\mu + D\omega_1 + E\bar{\omega}_1 + F\omega)$$

with

$$(59) \qquad\qquad A = -3i\bar{\mu} + 2\overline{\alpha_{1\bar{1}}},$$

$$(60) \qquad\qquad D - \frac{1}{2}C = 2i|\mu|^2 + \rho - \overline{\alpha_{1\bar{1}}}\mu - \alpha_{01},$$

$$(61) \qquad\qquad C \text{ is imaginary},$$

$$(62) \qquad\qquad E = i\mu^2 + \mu\alpha_{1\bar{1}} - \alpha_{0\bar{1}},$$

and

F is arbitrary.

PROOF. We have

$$d\Omega_1 - \frac{1}{2}\phi\Omega_1 = \frac{d\lambda}{\lambda}\Omega_1 + \lambda(\alpha_{ij}\omega_i\omega_j + d\mu\omega + i\mu\omega_1\bar{\omega}_1)$$
$$- \frac{1}{2}\left(\frac{d\lambda}{\lambda} + \frac{d\bar{\lambda}}{\bar{\lambda}} - i\bar{\mu}\omega_1 + i\mu\bar{\omega}_1 + 2\rho\omega\right)\Omega_1.$$

So

$$(63) \qquad d\Omega_1 - \frac{1}{2}\phi\Omega_1 = -\frac{1}{2}\left(\frac{d\lambda}{\lambda} - \frac{d\bar{\lambda}}{\bar{\lambda}} - i\bar{\mu}\omega_1 + i\mu\bar{\omega}_1 + 2\rho\omega\right)\Omega_1$$
$$+ \lambda(\alpha_{1\bar{1}}\omega_1\bar{\omega}_1 + \alpha_{01}\omega\omega_1 + \alpha_{0\bar{1}}\omega\bar{\omega}_1)$$
$$+ \lambda d\mu\omega + i\lambda\mu\omega_1\bar{\omega}_1.$$

We want the right-hand side to equal

$$\frac{1}{2}\psi\Omega_1 + \Omega_3\Omega,$$

so ψ and Ω_3 must have the form

$$\psi = \frac{d\lambda}{\lambda} - \frac{d\bar{\lambda}}{\bar{\lambda}} + A\omega_1 + B\bar{\omega}_1 + C\omega$$

and

$$\Omega_3 = \frac{1}{\lambda}(d\mu + D\omega_1 + E\bar{\omega}_1 + F\omega).$$

Equating the coefficients of $\omega_1\bar{\omega}_1$ we see that $B = -3i\mu - 2\alpha_{1\bar{1}}$. The condition $\psi = -\bar{\psi}$ then gives A:

$$A = -\bar{B} = -3i\bar{\mu} + 2\overline{\alpha_{1\bar{1}}}.$$

Looking at the coefficients of $\omega\omega_1$ we obtain

$$D - \frac{1}{2}C = \rho + 2i|\mu|^2 - \overline{\alpha_{1\bar{1}}}\mu - \alpha_{01},$$

while the coefficients of $\omega\bar{\omega}_1$ yield

$$E = i\mu^2 + \mu\alpha_{1\bar{1}} - \alpha_{0\bar{1}}.$$

Clearly F remains arbitrary.

We shall use Γ^0 and Γ^1 to denote functions of α_{ij} and their derivatives. Thus Γ^j is a constant on the fibers of B and is zero if $M = Q$.

LEMMA 8. ψ and Ω_3 satisfy

(64) $$d\psi = -3i\bar{\Omega}_3\Omega_1 - 3i\Omega_3\bar{\Omega}_1$$

if and only if

(65) $$C = -3i|\mu|^2 + \Gamma^0$$

and

(66) $$F = \bar{\mu}E + \mu\bar{D} + \frac{i}{3}A\alpha_{0\bar{1}} - \frac{i}{3}\bar{A}\overline{\alpha_{01}} + \Gamma^1,$$

with

(67) $$\Gamma^0 = \frac{3}{4}(\alpha_{01} - \overline{\alpha_{01}}) - \frac{i}{2}(\alpha_{1\bar{1},1} + \overline{\alpha_{1\bar{1},\bar{1}}}) + i|\alpha_{1\bar{1}}|^2$$

and

(68) $$\Gamma^1 = \frac{i}{3}\overline{\Gamma^0_1} + \frac{2}{3}i(\overline{\alpha_{1\bar{1}}}\alpha_{0\bar{1}} - \alpha_{1\bar{1}}\overline{\alpha_{01}} - \alpha_{1\bar{1},0}).$$

PROOF. We start with

$$\psi = \frac{d\lambda}{\lambda} - \frac{d\bar{\lambda}}{\bar{\lambda}} + A\omega_1 - \bar{A}\bar{\omega}_1 + C\omega$$

and

$$\Omega_3 = \frac{1}{\lambda}(d\mu + D\omega_1 + E\bar{\omega}_1 + F\omega),$$

with A and E given by (59) and (62), D and C related by (60), and F still arbitrary.

We want to compute $d\psi + 3i\overline{\Omega}_3\Omega_1 + 3i\Omega_3\overline{\Omega}_1$. We first compute the co-efficient of $\omega_1\overline{\omega}_1$ in each term. From $d\psi$ we have $iC + A\alpha_{1\overline{1}} + \overline{A}\overline{\alpha_{1\overline{1}}} - 2(\alpha_{1\overline{1},1} + \overline{\alpha_{1\overline{1}},\overline{1}})$; from $3i\overline{\Omega}_3\Omega_1$ we have $-3i\overline{D}$; and from $3i\Omega_3\overline{\Omega}_1$ we have $3iD$. Thus we set

$$iC + 3i(D - \overline{D}) = 2(\alpha_{1\overline{1},1} + \overline{\alpha_{1\overline{1}},1}) - A\alpha_{1\overline{1}} - \overline{A}\overline{\alpha_{1\overline{1}}}.$$

We combine this with the expression for $D - \frac{1}{2}C$ above to obtain

(69) $$C = -3i|\mu|^2 + \Gamma^0$$

with

(70) $$\Gamma^0 = \frac{3}{4}(\alpha_{01} - \overline{\alpha_{01}}) - \frac{i}{2}(\alpha_{1\overline{1},1} + \overline{\alpha_{1\overline{1}},\overline{1}}) + i|\alpha_{1\overline{1}}|^2.$$

That Γ^0 is indeed imaginary follows from Exercise 2.

Using the above expression for C we have

(71) $$D = \rho + \frac{i}{2}|\mu|^2 - \overline{\alpha_{1\overline{1}}}\mu - \alpha_{01} + \frac{1}{2}\Gamma^0.$$

Next we compute the coefficients of $\omega\omega_1$. From $d\psi$ we have $2\overline{\alpha_{1\overline{1}},0} + A\alpha_{01} - \overline{A}\overline{\alpha_{0\overline{1}}} - \Gamma^0_1$; from $3i\overline{\Omega}_3\Omega_1$ we have $3i(F - E\mu)$; and from $3i\Omega_3\overline{\Omega}_1$ we have $-3iD\overline{\mu}$. Thus we set

$$2\overline{\alpha_{1\overline{1}},0} + A\alpha_{01} - \overline{A}\overline{\alpha_{0\overline{1}}} - \Gamma^0_1 + 3i\overline{F} - 3i\overline{E}\mu - 3iD\overline{\mu} = 0.$$

This gives

(72) $$F = \overline{D}\mu + E\overline{\mu} + \mu\overline{\alpha_{0\overline{1}}} + \overline{\mu}\alpha_{0\overline{1}} + \Gamma^1$$

with

(73) $$\Gamma^1 = \frac{i}{3}\overline{\Gamma^0_1} + \frac{2}{3}i(\overline{\alpha_{1\overline{1}}}\alpha_{0\overline{1}} - \alpha_{1\overline{1}}\overline{\alpha_{0\overline{1}}} - \alpha_{1\overline{1},0}).$$

Now $d\psi + 3i\overline{\Omega}_3\Omega_1 + 3i\Omega_3\overline{\Omega}_1$ clearly contains no $d\lambda$, $d\overline{\lambda}$, or $d\rho$ terms. Further, if $d\mu$ does not occur then also $d\overline{\mu}$ is absent. So to complete the proof of the lemma we only need show that $d\psi + 3i\overline{\Omega}_3\Omega_1 + 3i\Omega_3\overline{\Omega}_1$ cannot contain any $d\mu$ term. But this is clear since $d\psi$ contributes $-3id\mu\overline{\omega}_1$ while $3i\Omega_3\overline{\Omega}_1$ contributes $3id\mu\overline{\omega}_1$.

The forms Ω, Ω_1, ϕ, ψ, Ω_3 are uniquely defined. Set $\Omega_2 = \frac{1}{2}(\phi + \psi)$. We have the equations

(74) $$d\Omega = i\Omega_1\overline{\Omega}_1 + \phi\Omega,$$

(75) $$d\Omega_1 = \Omega_2\Omega_1 + \Omega_3\Omega,$$

(76) $$d\psi = -3i\overline{\Omega}_3\Omega_1 - 3i\Omega_3\overline{\Omega}_1.$$

LEMMA 9. $(d\phi - i\Omega_3\overline{\Omega}_1 - i\Omega_1\overline{\Omega}_3)\Omega = 0$.

PROOF. Taking the exterior derivative of $d\Omega$ we obtain

$$0 = (d\phi)\Omega - i\phi\Omega_1\overline{\Omega}_1 + i\Omega_2\Omega_1\overline{\Omega}_1 + i\overline{\Omega}_2\Omega_1\overline{\Omega}_1 + i\Omega_3\Omega\overline{\Omega}_1 + i\overline{\Omega}_3\Omega_1\Omega$$
$$= (d\phi - i\Omega_3\overline{\Omega}_1 - i\Omega_1\overline{\Omega}_3)\Omega + i\Omega_1\overline{\Omega}_1(-\phi + \Omega_2 + \overline{\Omega}_2).$$

But $\overline{\Omega}_2 = \frac{1}{2}(\phi - \psi)$ and so the last term is zero.

EXERCISE 9. Use $dd\Omega_1 = 0$ to derive that

$$d\psi\Omega_1\Omega = 3i\Omega_3\Omega_1\overline{\Omega}_1\Omega.$$

We may define a form Ω_4, which is unique up to a term in Ω, so that

$$d\phi = i\Omega_3\overline{\Omega}_1 + i\Omega_1\overline{\Omega}_3 + 2\Omega_4\Omega.$$

We may restrict the term in Ω to be real and then Ω_4 is real. A simple computation shows that

(77) $$\Omega_4 = \frac{1}{|\lambda|^2}\left\{dp + G\omega + \frac{i\mu}{2}d\overline{\mu} - \frac{i\overline{\mu}}{2}d\mu + a\omega_1 + \overline{a}\,\overline{\omega}_1\right\},$$

where G is a real function not yet determined and

(78) $$a = \frac{1}{2}(i\mu\overline{\alpha_{10}} - i\overline{\mu}\alpha_{10} - i\overline{\mu}D - iF + i\mu\overline{E}).$$

We shall determine G by requiring that a certain term in $d\Omega_3 + \Omega_1\Omega_4 + \overline{\Omega}_2\Omega_3$ is zero. As with $d\phi$, we have that many terms in this 2-form are automatically zero.

LEMMA 10. *If Ω_4 is any real form such that*

(79) $$d\phi = i\Omega_3\overline{\Omega}_1 + i\Omega_1\overline{\Omega}_3 + 2\Omega_4\Omega,$$

then

(80) $$d\Omega_3 + \Omega_1\Omega_4 + \overline{\Omega}_2\Omega_3 = \left(h\Omega_1 + \frac{r}{\overline{\lambda}^3\lambda}\overline{\Omega}_1\right)\Omega,$$

where h and r are real and r is a function only of $x \in M$.

PROOF. Compute $dd\Omega_2$ and use for $d\Omega_3$ the expression $d\Omega_3 = -\Omega_1\Omega_4 - \overline{\Omega}_2\Omega_3 + R$. The result can be written

$$2i\Omega_1\overline{R} + i\overline{\Omega}_1 R = \Omega(-i\Omega_3\overline{\Omega}_3 + \phi\Omega_4 - d\Omega_4).$$

Thus $R = S_1\Omega$. Further, the right-hand side of the above equation is real; so

$$2i\Omega_1\overline{R} + i\overline{\Omega}_1 R = -2i\overline{\Omega}_1 R - i\Omega_1\overline{R}.$$

Thus $i\overline{\Omega}_1 R$ is real and so we also have that $R = S_2\Omega_1 + S_3\overline{\Omega}_1$. It follows that $R = h\Omega_1 + \beta\overline{\Omega}_1$ with h real. We now determine the form of β. We take the exterior derivative of

$$d\Omega_3 = -\Omega_1\Omega_4 - \overline{\Omega}_2\Omega_3 + (h\Omega_1 + \beta\overline{\Omega}_1)\Omega$$

and then wedge with Ω_1 to obtain

$$d\beta\,\Omega\Omega_1\overline{\Omega}_1 + 3\beta\overline{\Omega}_2\Omega\Omega_1\overline{\Omega}_1 + \beta\Omega_2\Omega\Omega_1\overline{\Omega}_1 = 0.$$

Thus $d\beta$ does not contain $d\mu$ or $d\rho$ and we may write

$$d\beta = \beta_0\Omega + \beta_1\Omega_1 + \beta_{\overline{1}}\overline{\Omega}_1 + \beta_2\Omega_2 + \beta_{\overline{2}}\overline{\Omega}_2.$$

Thus

$$\beta_2 + \beta = 0, \quad \beta_{\overline{2}} + 3\beta = 0,$$

and, as we see by equating the coefficients of $d\lambda$,

$$\beta_2 = \lambda\frac{\partial\beta}{\partial\lambda}.$$

Also

$$\beta_{\overline{1}} = \overline{\lambda}\frac{\partial\beta}{\partial\overline{\lambda}}.$$

We now show that this implies β has the desired special form. This would complete the proof of Lemma 10.

LEMMA 11. *Let* $\beta = \beta(x, \lambda, \overline{\lambda})$ *be a* C^∞ *function that satisfies*

$$\lambda\frac{\partial\beta}{\partial\lambda} + \beta = 0 \quad and \quad \overline{\lambda}\frac{\partial\beta}{\partial\overline{\lambda}} + 3\beta = 0.$$

Then $\beta = (\overline{\lambda}^3\lambda)^{-1}r(x)$.

PROOF. The expression for β was derived by formally integrating the two equations. This would only establish the result under an assumption such as $\beta(x, \lambda, \overline{\tau})$ is holomorphic in λ and τ. The general result, even for β only a C^1 function, comes from simply computing $\frac{\partial}{\partial\lambda}(\overline{\lambda}^3\lambda\beta)$ and $\frac{\partial}{\partial\overline{\lambda}}(\overline{\lambda}^3\lambda\beta)$.

Lemma 10 shows that no matter what real value we choose for G we have that (80) holds where h is some real quantity. A simple computation shows that the coefficient of $\omega\omega_1$ in $d\Omega_3 + \Omega_1\Omega_4 + \overline{\Omega}_2\Omega_3$ is

$$(81) \quad \frac{1}{\lambda}\left(D_0 - F_1 + \alpha_{01}D + \overline{\alpha_{01}}E - \frac{a\mu}{2} + \frac{AF}{2} - \frac{CD}{2} + \frac{i\overline{\mu}F}{2} + \rho D - G\right).$$

This is known to be real (but the computation to verify this would not be easy) and so there is a unique choice of G that makes this coefficient and hence h equal to zero. With this choice we have that Ω, Ω_1, Ω_2, Ω_3, and Ω_4 are uniquely defined forms on the bundle B.

The equation for $d\Omega_4$ can be easily derived by taking the exterior derivatives of the equations for $d\Omega_2$ and $d\Omega_3$.

EXERCISE 10. Show

$$
(82) \qquad d\Omega_4 = i\Omega_3\overline{\Omega}_3 - (\Omega_2 + \overline{\Omega}_2)\Omega_4 - S\Omega\Omega_1 - \overline{S}\Omega\overline{\Omega}_1 ,
$$

where S is some function on B with the property that $S \equiv 0$ on any set where $R \equiv 0$. (Hint: Cf. the remarks after (48)).

Let us summarize the main result of this construction. Each strictly pseudoconvex CR structure on M^3 defines an eight-dimensional sub-bundle B of some frame bundle. On B there are eight independent real 1-forms given by $\Omega, \Omega_1, \Omega_2, \Omega_3, \Omega_4$ with Ω and Ω_4 real, which satisfy

$$
\begin{cases}
d\Omega \ = i\Omega_1\overline{\Omega}_1 - \Omega(\Omega_2 + \overline{\Omega}_2), \\
d\Omega_1 = -\Omega_1\Omega_2 - \Omega\Omega_3 , \\
d\Omega_2 = 2i\Omega_1\overline{\Omega}_3 + i\overline{\Omega}_1\Omega_3 - \Omega\Omega_4 , \\
d\Omega_3 = -\Omega_1\Omega_4 - \overline{\Omega}_2\Omega_3 - R\Omega\overline{\Omega}_1 , \\
d\Omega_4 = i\Omega_3\overline{\Omega}_3 - (\Omega_2 + \overline{\Omega}_2)\Omega_4 + (S\Omega_1 + \overline{S}\overline{\Omega}_1)\Omega.
\end{cases}
$$

Any choice of 1-forms ω and ω_1 that define the CR structure of M and that satisfy $d\omega = i\omega_1\overline{\omega}_1$ leads to local coordinates x, λ, μ, ρ on B. In these coordinates $R = \frac{r(x)}{\overline{\lambda}^3\lambda}$ where r is some function on M. Further, if $R \equiv 0$ on some open set then $S \equiv 0$ on the same set.

It might be useful to collect here the explicit definitions of these forms in local coordinates:

$$
(83) \qquad \Omega \ = |\lambda|^2\omega ,
$$

$$
(84) \qquad \Omega_1 = \lambda(\omega_1 + \mu\omega) ,
$$

$$
(85) \qquad \Omega_2 = \frac{1}{2}(\phi + \psi) = \frac{d\lambda}{\lambda} + (-2i\overline{\mu} + \overline{\alpha_{1\overline{1}}})\omega_1 + (-i\mu - \alpha_{1\overline{1}})\overline{\omega}_1 ,
$$

$$
\left(-\frac{3}{2}i|\mu|^2 + \rho - \Gamma^0\right)\omega ,
$$

$$
(86) \qquad \Omega_3 = \frac{1}{\overline{\lambda}}\left(d\mu + D\omega_1 + E\overline{\omega}_1 + F\omega\right) ,
$$

$$
(87) \qquad \Omega_4 = \frac{1}{|\lambda|^2}\left(d\rho + G\omega + \frac{i\mu}{2}d\mu - \frac{i\overline{\mu}}{2}d\overline{\mu} + a\omega_1 + \overline{a}\,\overline{\omega}_1\right) ,
$$

where

$$d\omega_1 = \alpha_{01}\omega\omega_1 + \alpha_{0\bar{1}}\omega\overline{\omega}_1 + \alpha_{1\bar{1}}\overline{\omega}_1,$$

$$D = \rho + \frac{i}{2}|\mu|^2 - \overline{\alpha_{1\bar{1}}}\mu - \alpha_{01} + \frac{1}{2}\Gamma^0,$$

$$E = i\mu^2 + \mu\alpha_{1\bar{1}} - \alpha_{0\bar{1}},$$

$$F = \overline{\mu}E + \mu\overline{D} + \mu\overline{\alpha_{01}} + \overline{\mu}\alpha_{0\bar{1}} + \Gamma^1,$$

$$\Gamma^0 = -\frac{i}{2}(\overline{\alpha_{1\bar{1}}}_{,\bar{1}} + \alpha_{1\bar{1},1}) + i|\alpha_{1\bar{1}}|^2 + \frac{3}{4}(\alpha_{01} - \overline{\alpha_{01}}),$$

$$\Gamma^1 = \frac{i}{3}\overline{\Gamma^0_1} + \frac{2}{3}i(\overline{\alpha_{1\bar{1}}}\alpha_{0\bar{1}} - \alpha_{1\bar{1}}\overline{\alpha_{01}} - \alpha_{1\bar{1},0}),$$

$$a = \frac{1}{2}(i\mu\overline{\alpha_{\bar{1}0}} - i\overline{\mu}\alpha_{10} - i\overline{\mu}D - iF + i\mu\overline{E}),$$

$$G = \rho D + D_0 - F_1 + \alpha_{01}D + \overline{\alpha_{0\bar{1}}}E - \frac{a\mu}{2} + \frac{AF}{2} - \frac{CD}{2} + \frac{i\overline{\mu}}{2}F.$$

We now have two constructions of what we may call a CR connection on M^3. Several other constructions of presumably isomorphic connections are in the literature (e.g., [Tan2], [CM], [BDS]) which apply also to higher-dimensional CR structures (but always under the assumption that M is non-degenerate and of hypersurface type). We want to outline a specialization of [CM] to dimension three in order to show the variety of approaches to the basic construction and also to provide a guide to the higher-dimensional results. Again, as with all modern treatments, the bundle appears at the beginning. Do not forget, however, that many results follow directly and rigorously from Cartan's construction even without clearly specifying the bundle in which the geometry is defined. We follow [CM, pp. 250–259] with notation specialized to the case of $n = 1$ (where dim $M = 2n + 1$) and start with a CR structure given by $\{\theta, \theta^1\}$. We will need to admit some first-order information about θ. In the approach just given this was done by including the 0-jet of θ_1 but a certain 1-jet of θ. Here is another way. Let θ be some local choice for the annihilator of H, chosen such that for some θ^1 (and hence all θ^1)

$$d\theta = ig\theta^1 \wedge \overline{\theta^1} \quad (\mathrm{mod}\ \theta),$$

with $g > 0$. Consider the half-line bundle E over M defined by

$$E = \{(x, u\theta) \mid x \in M, \ u > 0\}.$$

The form ω given by

$$\omega(T) = u\theta(\pi_* T), \quad \text{for } T \in TE_{(x, u\theta)},$$

is clearly well-defined on E and, in particular, does not depend on the choice of θ. Note that, with only slight abuse of notation, we may write

$$\omega = u\theta,$$

where θ is the 1-form on M and ω is this well-defined 1-form on E. Let us now consider all sets of 1-forms on E,

$$(\omega, \omega^1, \overline{\omega^1}, \phi),$$

such that

1. ω is the above 1-form,
2. ω^1 is the pull-back under projection of some multiple of θ^1,
3. ϕ is a real 1-form on E,
4. $d\omega = i\omega^1\overline{\omega^1} + \omega\phi$.

The set Y of all such $(\omega, \omega^1, \overline{\omega^1}, \phi)$ is a principle fiber bundle over E with group G_1 given by the matrices

$$\begin{pmatrix} 1 & 0 & 0 & 0 \\ v & u_1^1 & 0 & 0 \\ \overline{v} & 0 & \overline{u_1^1} & 0 \\ s & iu_1^1\overline{v} & -iuu_1^1v & 1 \end{pmatrix},$$

with $|u_1^1| = 1$ and s real. Thus dim $Y = 8$. Again, there are well-defined forms on Y which we may denote by ω, ω^1, $\overline{\omega^1}$, and ϕ.

Let us see what these forms look like in local coordinates. We start with the form θ on M and the form $\omega = u\theta$ on E.

$$d\omega = du\theta + ud\theta$$
$$= \frac{du}{u}\omega + u(i\sigma^1\overline{\sigma^1} + \theta\widetilde{\phi}),$$

for some $\widetilde{\phi}$ and σ^1. Set $\theta^1 = \sqrt{u}\sigma^1$. Then

$$d\omega = i\theta^1\overline{\theta^1} + \omega\left(-\frac{du}{u} + \widetilde{\phi}\right).$$

Introduce coordinates on Y by

$$(x, u, u_1^1, v, s) \longrightarrow (x, \omega, \omega^1, \phi),$$

where

$$\omega = u\theta,$$
$$\omega^1 = u_1^1\theta^1 + v\theta,$$
$$\phi = -\frac{du}{u} + \widetilde{\phi} + s\theta + iu_1^1\overline{v}\,\theta^1 - i\overline{u_1^1}\,v\,\overline{\theta^1}.$$

So we indeed have

$$d\omega = i\omega^1\overline{\omega^1} + \omega\phi.$$

Note that $\{\omega, \omega^1, \overline{\omega^1}, \phi\}$ are linearly independent. Also note that

(88) $$d\omega^1 \equiv 0 \bmod \{\omega, \omega^1\}.$$

This is as forms on Y; for forms on M the relation, which is trivial, is

$$d\theta^1 \equiv 0 \bmod \{\theta, \theta^1\}.$$

We have an 8-dimensional bundle Y and forms $\omega, \omega^1, \overline{\omega^1}$, and ϕ with

$$d\omega = i\omega^1\overline{\omega^1} + \omega\phi.$$

THEOREM [**CM**, p. 258]. *There are four more uniquely determined forms* $\phi_1^1, \phi^1, \overline{\phi^1}$, *and* ψ *such that the forms*

$$\omega, \qquad \omega^1, \qquad \overline{\omega^1}, \qquad \phi, \qquad \phi_1^1, \qquad \phi^1, \qquad \overline{\phi^1}, \qquad \psi$$

are linearly independent and satisfy

$$d\omega = i\omega^1\overline{\omega^1} + \omega(\phi_1^1 + \overline{\phi_1^1}),$$
$$d\omega^1 = \omega^1\phi_1^1 + \omega\phi^1,$$
$$d\phi^1 = (\phi_1^1 + \overline{\phi_1^1})\phi^1 + \phi^1\phi_1^1 - \frac{1}{2}\psi\omega^1 + Q\overline{\omega_1}\omega,$$
$$d\psi = (\phi_1^1 + \overline{\phi_1^1})\psi + (R\omega^1 + \overline{R}\overline{\omega^1})\omega + 2i\phi^1\overline{\phi^1},$$
$$d\phi_1^1 = i\overline{\omega^1}\phi^1 - 2i\overline{\phi^1}\omega^1 - \frac{1}{2}\psi\omega.$$

This completely solves the equivalence problem in the same way that Cartan's construction did. Actually this approach just redoes Cartan in a way suitable for the higher-dimensional extension. The isomorphism between the two results is

$$\omega = \Omega,$$
$$\omega^1 = \Omega_1,$$
$$\phi_1^1 = -\Omega_2,$$
$$\phi^1 = -\Omega_3,$$
$$\psi = 2\Omega_4,$$
$$Q = -R,$$
$$R = 2S.$$

We start the proof of this theorem by rewriting (88) as

(89) $$d\omega^1 = \omega^1\phi_1^1 + \omega\phi^1.$$

Of course, ϕ_1^1 and ϕ^1 are not uniquely determined and we look for auxiliary conditions to impose to ensure uniqueness.

LEMMA 11 [Lemma 4.1 of [**CM**]]. *There exists some* ϕ_1^1 *that satisfies both*

    1) $d\omega^1 = \omega^1\phi_1^1 + \omega\phi^1$

and

    2) $2\operatorname{Re}\phi_1^1 = \phi$.

Further, any other solution to these equations is given by $\phi_1^1 + i\sigma\omega$ *for an arbitrary real function* σ.

The proof is easy. We pick some particular 1-forms satisfying

$$d\omega^1 = \omega^1\alpha + \omega\beta.$$

Then all possible solutions to 1) and 2) are given by

$$\phi_1^1 = \alpha + \lambda_0\omega + \lambda_1\omega^1$$

and

$$\phi^1 = \beta + \lambda_0\omega^1 + \lambda_2\omega,$$

with λ_0, λ_1, λ_2 arbitrary. Take the exterior derivative of both sides of

(90) $$d\omega = i\omega^1\overline{\omega^1} + \omega\phi,$$

and use (89) to obtain

$$i\omega^1\overline{\omega^1}(-\phi_1^1 - \overline{\phi_1^1} + \phi) - i\omega\overline{\omega^1}(\phi^1 + \overline{\phi^1}) - \omega\,d\phi = 0.$$

This implies

(91) $$2\,\mathrm{Re}\,\phi_1^1 \equiv \phi \bmod \{\omega, \omega^1, \overline{\omega^1}\}.$$

LEMMA 12. *If for two 1-forms* γ *and* δ *on* Y, δ *real, one has* $2\,\mathrm{Re}\,\gamma \equiv \delta \bmod \{\omega, \omega^1, \overline{\omega^1}\}$ *then for a unique complex-valued function* λ, *a unique real-valued function* ξ, *and an arbitrary real-valued function* η *one has*

$$2\,\mathrm{Re}\,(\gamma + (\xi + i\eta)\omega + \lambda\omega^1) = \delta.$$

PROOF. We are given that there exist σ_0, σ_1, $\sigma_{\bar{1}}$ with

$$2\,\mathrm{Re}\,\gamma = \delta + \sigma_0\omega + \sigma_1\omega^1 + \sigma_{\bar{1}}\overline{\omega^1}.$$

Since δ is real, this implies σ_0 is real and $\overline{\sigma_1} = \sigma_{\bar{1}}$. Thus $2\,\mathrm{Re}\,(\gamma - \frac{1}{2}\sigma_0\omega - \sigma_1\omega^1) = \delta$. So $\xi = -\frac{1}{2}\sigma_0$, $\lambda = \sigma_1$, and η is arbitrary.

Apply this lemma to (91) to prove Lemma 11.

We shall use the following lemma to determine the remaining coefficient in ϕ_1^1.

LEMMA 13 (Lemma 4.2 of [CM]). *If* Φ *is a 2-form with*

$$\Phi + \overline{\Phi} \equiv 0 \bmod \omega \quad and \quad \omega^1 \wedge \Phi \equiv 0 \bmod \omega,$$

then $\Phi \equiv h\omega^1\overline{\omega^1} \bmod \omega$, *where* h *is a real-valued function.*

PROOF. The two conditions imply at first that

$$\Phi = \omega^1 \wedge S + \omega \wedge T,$$

where $S \in \{\omega, \omega^1, \overline{\omega^1}\}$. The first condition further implies that $S = \alpha\omega + \beta\omega^1 + \gamma\overline{\omega^1}$ with γ real. So we are done.

LEMMA 14. *The condition*

$$(92) \qquad d\phi_1^1 - i\overline{\omega^1}\phi^1 + 2i\overline{\phi^1}\omega^1 \equiv 0 \bmod \omega$$

determines ϕ_1^1 *completely and* ϕ^1 *up to a multiple of* ω.

PROOF. Take the exterior derivative of

$$d\omega = i\omega^1\overline{\omega^1} + \omega\phi,$$

and use $2\operatorname{Re}\phi_1^1 = \phi$ to obtain

$$(93) \qquad \omega(d\phi + i\overline{\omega^1}\phi^1 - i\omega^1\overline{\phi^1}) = 0.$$

Set

$$\Phi = d\phi_1^1 - i\overline{\omega^1}\phi^1 + 2i\overline{\phi^1}\omega^1.$$

So

$$\Phi + \overline{\Phi} = d\phi + i\overline{\omega^1}\phi^1 + i\overline{\phi^1}\omega^1 \equiv 0 \bmod \omega.$$

Now take the exterior derivative of

$$d\omega^1 = \omega^1\phi_1^1 + \omega\phi^1$$

to obtain

$$(94) \qquad \omega^1\Phi \equiv 0 \bmod \omega.$$

Thus, by the previous lemma, we have

$$(95) \qquad d\phi_1^1 - i\overline{\omega^1}\phi^1 - 2i\omega^1\overline{\phi^1} \equiv h\omega^1\overline{\omega^1} \bmod \omega.$$

It follows from (89) and Lemma 11 that

$$\phi_1^1 = \alpha + i\sigma\omega$$

and

$$\phi^1 = \beta + i\sigma\omega^1 + \lambda\omega,$$

where α and β may be considered as known forms and σ and λ are arbitrary functions with σ real. It is easily seen that there is a unique σ that achieves

$$(96) \qquad d\phi_1^1 - i\overline{\omega^1}\phi^1 + 2i\overline{\phi^1}\omega^1 \equiv 0 \bmod \omega.$$

That σ is real follows from the observation that (95) holds for all σ and, in particular, for $\sigma = 0$.

Now we turn to the ω coefficient in ϕ^1.

LEMMA 15 (Lemma 4.4 of [CM]). *The condition*

$$d\phi^1 - \phi\phi^1 - \phi^1\phi_1^1 + \frac{1}{2}\psi\omega^1 \equiv 0 \bmod \omega$$

determines ϕ^1 *uniquely.*

PROOF. We rewrite (96) as

(97) $$d\phi_1^1 - i\overline{\omega^1}\phi^1 + 2i\overline{\phi^1}\omega^1 = \lambda\omega,$$

for some 1-form λ, and rewrite (93) as

(98) $$d\phi + i\overline{\omega^1}\phi^1 - i\omega^1\overline{\phi^1} = \omega\psi,$$

for some 1-form ψ. We return to the exterior derivative of $d\omega^1$ and this time write out our full conclusion:

$$(d\phi_1^1 - i\overline{\omega^1}\phi^1 + 2i\overline{\phi^1}\omega^1)\omega^1 + (d\phi^1 - \phi\phi^1 - \phi^1\phi_1^1)\omega = 0.$$

Note that this first term is just $\lambda\omega\omega^1$. Thus

(99) $$d\phi^1 - \phi\phi^1 - \phi^1\phi_1^1 - \lambda\omega^1 = \mu\omega,$$

for some 1-form μ.

We add to (97) its conjugate (recalling Lemma 11). The result is

$$d\phi + i\overline{\omega^1}\phi^1 - i\omega^1\overline{\phi^1} = (\lambda + \bar{\lambda})\omega.$$

Comparing this to (98), we see that

(100) $$\lambda + \bar{\lambda} \equiv -\psi \bmod \omega.$$

But we can do better than this. We have not yet taken the exterior derivative of (97). Let us do so now, working modulo ω. So

$$-i\overline{\omega^1}\phi_1^1\phi^1 + i\overline{\omega^1}d\phi^1 + 2i\overline{d\phi^1}\omega^1$$
$$- 2i\overline{\phi^1}\omega^1\phi_1^1 \equiv -\lambda i\omega^1\overline{\omega^1}.$$

Let us substitute for $d\phi^1$ by using (99). This yields

$$\overline{\omega^1}(i\lambda\omega^1 - \overline{\phi_1^1}\phi^1 + i\phi^1\phi_1^1 + \phi\phi^1)$$
$$+ 2i\omega^1(\overline{\lambda\omega^1} - \overline{\phi_1^1}\overline{\phi^1} + \overline{\phi^1}\phi_1^1 + \overline{\phi\phi^1})$$
$$= -\lambda i\omega^1\overline{\omega^1}.$$

Now ϕ_1^1 has already been completely determined and ϕ^1 has been completely determined modulo ω. It follows that each term above is completely determined modulo ω and it is meaningful to equate the coefficients of $\omega^1\overline{\omega^1}$. This yields

$$(\lambda - \bar{\lambda})\omega^1\overline{\omega^1} = 0.$$

Thus using (100) we have

(101)
$$\lambda \equiv -\frac{1}{2}\psi \bmod \{\omega, \omega^1, \overline{\omega^1}\},$$

which we immediately rewrite as

$$\lambda = -\frac{1}{2}\psi + V\omega^1 + W\overline{\omega^1} + a\omega,$$

and substitute into (99). Thus

(102)
$$d\phi^1 - \phi\phi^1 - \phi^1\phi_1^1 + \frac{1}{2}\psi\omega^1 = -W\omega^1\overline{\omega^1} + \nu\omega,$$

for some 1-form ν. Recall that in all this ϕ^1 is only determined up to a multiple of ω. So if we fix one choice for ϕ^1, say α, then we have from (102)

$$d\alpha - \phi\alpha - \alpha\phi_1^1 + \frac{1}{2}\psi\omega^1 = -W\omega^1\overline{\omega^1} + \nu\omega$$

and

$$d\phi^1 - \phi\phi^1 - \phi^1\phi_1^1 + \frac{1}{2}\psi\omega^1 = -W\omega^1\overline{\omega^1} + \tilde{\nu}\omega - i\sigma\omega^1\overline{\omega^1},$$

where $\tilde{\nu}$ depends on the function σ. But it is clear that there is a unique σ that yields

(103)
$$d\phi^1 - \phi\phi^1 - \phi^1\phi_1^1 + \frac{1}{2}\psi\omega^1 \equiv 0 \bmod \omega.$$

This concludes the proof of Lemma 15.

We now have on Y the well-defined forms: $\omega, \omega^1, \overline{\omega^1}, \phi, \phi_1^1, \phi^1$, $\overline{\phi^1}$. It is easily checked by introducing local coordinates that these forms are linearly independent. We need one more form. The obvious candidate is ψ from (98). It can be shown that ψ is linearly independent of the other seven forms. Clearly it is also unique except for a multiple of ω which we now want to fix.

LEMMA 16 (Lemma 4.5 of [CM]). *The form ψ is completely determined by the additional condition*

$$d\psi - \phi\psi - 2i\phi^1\overline{\phi^1} \equiv 0 \bmod \omega.$$

PROOF. We start by rewriting (103) as

(104)
$$d\phi^1 - \phi\phi^1 - \phi^1\phi_1^1 + \frac{1}{2}\psi\omega^1 = \nu\omega,$$

where ν is determined up to a multiple of ω. Next we differentiate (98), use (104), and look only at terms containing ω. This yields

(105)
$$d\psi - \phi\psi - 2i\phi^1\overline{\phi^1} \equiv -i\omega^1\overline{\nu} - i\nu\overline{\omega^1} \pmod{\omega}.$$

In order to see the nature of the 1-form ν, we differentiate (104) working mod ω and keeping only terms with $\omega^1\overline{\omega^1}$. The result is

$$\nu\omega^1\overline{\omega^1} \equiv 0 \bmod \omega.$$

Thus
$$\nu = P\omega^1 + Q\overline{\omega^1} + R\omega.$$

We use this to rewrite (105) as
$$d\psi \equiv \phi\psi + 2i\phi^1\overline{\phi^1} - i(P + \overline{P})\omega^1\overline{\omega^1} \mod \omega.$$

Now ψ was well-defined up to a real multiple of ω and by our usual argument this multiple may be made unique by requiring that $P + \overline{P} = 0$. This ends our search. We now have the eight 1-forms on Y.

CHAPTER 7

Geometric Consequences

We want to derive various geometric results from Cartan's solution to the equivalence problem for CR structures. We start by making more explicit the fact that Cartan's construction is CR invariant. Recall that a map $\phi\colon M \to \widetilde{M}$ of CR manifolds is a CR map if it satisfies the equivalent set of conditions:

(i) $\phi_* H \subset \widetilde{H}$ and $\phi_* J v = J \phi_* v$, for all $v \in H$,

or

(ii) $\phi_* V \subset \widetilde{V}$ where V is defined by $JV = -iV$,

or, for $\dim M = 3$,

(iii) $\phi^*(\tilde{\omega}) = r\omega$ and $\phi^*(\tilde{\omega}_1) = \lambda(\omega_1 + \mu\omega)$ where $\{\omega, \omega_1\} = V^\perp$ and is real.

Consider the one-forms $\Omega, \Omega_1, \Omega_2, \Omega_3, \Omega_4$. For notational convenience let us set $\Omega_0 = \Omega$.

LEMMA 1. *Any CR diffeomorphism* $\phi\colon M \to \widetilde{M}$ *lifts to a fiber-preserving diffeomorphism* $\Phi\colon B \to \widetilde{B}$ *that satisfies* $\Phi^* \widetilde{\Omega}_j = \Omega_j$, $j = 0, 1, 2, 3, 4$.

PROOF. From one point of view there is nothing to prove. The CR diffeomorphism allows us to identify the two CR structures. Cartan's construction of the differential forms depends only on the CR structure. From this point of view Φ is actually the identity. We also have the following exercise for readers who might prefer a more concrete argument. Here we use one of the constructions that make the bundle explicit.

EXERCISE 1. (a) If $\phi\colon U \to V$ is a diffeomorphism then $d(\phi^*\sigma) = \phi^* d\sigma$ for any differential form on V. (Hint: It suffices to prove this for functions and one-forms.)

(b) Let $\phi\colon M \to \widetilde{M}$ be a CR diffeomorphism. Choose bases $\{\omega, \omega_1\}$ and $\{\tilde{\omega}, \tilde{\omega}_1\}$ for the CR structures with $d\omega = i\omega_1\overline{\omega}_1$ and $d\tilde{\omega} = i\tilde{\omega}_1\overline{\tilde{\omega}}_1$. Set $\phi^*(\tilde{\omega}) = r\omega$ and $\phi^*(\tilde{\omega}_1) = \alpha(\omega_1 + \beta\omega)$. Show

$$r = |\alpha|^2 \quad \text{and} \quad dr = r_0\omega + i\overline{\beta}|\alpha|^2\omega_1 - i\beta|\alpha|^2\overline{\omega}_1$$

for some function r_0.

(c) Recall that local coordinates $\{\lambda, \mu, \rho\}$ may be introduced on B by taking as points of the fiber of B above $x \in M$

$$j_x^1(R\omega) \times j^0(\lambda(\omega_1 + \mu\omega)),$$

where at x, $R = |\lambda|^2$ and

$$dR = -2|\lambda|^2 \rho\omega + i|\lambda|^2 \overline{\mu}\omega_1 - i|\lambda|^2 \mu\overline{\omega}_1.$$

Define a map $\Psi: \widetilde{B} \to B$ by trying to set

$$\Psi(j^1(\widetilde{R}\widetilde{\omega}) \times j^0(\widetilde{\lambda}(\widetilde{\omega}_1 + \widetilde{\mu}\widetilde{\omega})))$$
$$= j^1((\widetilde{R} \circ \phi)\phi^* \widetilde{\omega}) \times j \circ (\phi^*(\widetilde{\lambda}(\widetilde{\omega}_1 + \widetilde{\mu}\widetilde{\omega}))).$$

Show that the right-hand side is actually in B and that it is given in terms of the local coordinates by

$$\lambda = \widetilde{\lambda}\alpha,$$
$$\mu = \widetilde{\mu}r\alpha^{-1} + \beta,$$
$$\rho = -\frac{1}{2} r_0 r^{-1} + \widetilde{\rho}r - \frac{i}{2}\overline{\widetilde{\mu}}\alpha\beta + \frac{i}{2}\widetilde{\mu}\overline{\alpha\beta}.$$

(d) Show that $\Psi: \widetilde{B} \to B$ depends at each point in the fiber above $\phi(x)$ only on ϕ and its first and second derivations at x.

(e) For the map $\Phi: B \to \widetilde{B}$ given by $\Phi = \Psi^{-1}$ show that $\Phi^*(\widetilde{\Omega}) = \Omega$ and $\Phi^*(\widetilde{\Omega}_1) = \Omega_1$.

(f) Use the uniqueness result for Cartan's construction to conclude that also $\Phi^*(\widetilde{\Omega}_j) = \Omega_j$ for $j = 2, 3$, and 4.

Lemma 1 and (d) of the exercise have a simple consequence which already illustrates the usefulness of Cartan's construction.

EXERCISE 2. (a) Let M^3 be a nondegenerate hypersurface in $\mathbb{C}^2$ containing the origin. Let $\phi: U \to V$ be a biholomorphism of open connected neighborhoods of the origin with $\phi(0) = 0$. If $\phi(M \cap U) \subset M \cap V$ and $\phi(\zeta) = \zeta + O(|\zeta|^3)$ then ϕ is the identity map. (Hint: Show $\Phi: B \to B$ preserves some point in the fiber above the origin and thus must be the identity map.)

(b) Explicitly verify this result when M^3 is a piece of the hyperquadric Q.

We now turn to the converse of Lemma 1.

LEMMA 2. *Let* $\Phi: B \to \widetilde{B}$ *be any map with* $\Phi^*(\widetilde{\Omega}_j) = \Omega_j$, $j = 0, \ldots, 4$. *Then there is a CR diffeomorphism* $\phi: M \to \widetilde{M}$ *that has* Φ *as its lift.*

PROOF. Note that Φ is a diffeomorphism. Let us show that it maps the fibers of B to those of $\widetilde{B}$. First note that the fibers of B are the integral submanifolds for the closed differential ideal $\{\Omega, \Omega_1, \overline{\Omega}_1\}$. So let F_p be the fiber above $p \in M$. For each $v \in T(\Phi(F_p))$ there is a $u \in T(F_p)$ with

$\Phi_* u = v$. Thus $\widetilde{\Omega}_j v = \Omega_j \Phi_* u = (\Phi^* \widetilde{\Omega}_j) u = \Omega_j u$. Hence $\widetilde{\Omega}_0 v = \widetilde{\Omega}_1 v = 0$ and so $\Phi(F_p)$ is a fiber of $\widetilde{B}$ and Φ defines a map $\phi \colon M \to \widetilde{M}$. Now $\widetilde{\Omega} = |\tilde{\lambda}|^2 \tilde{\omega}$ and so

$$\Phi^* \widetilde{\Omega} = |\tilde{\lambda}|^2 \Phi^* \tilde{\omega} = |\tilde{\lambda}|^2 \phi^* \tilde{\omega},$$

and also

$$\Phi^* \widetilde{\Omega} = \Omega = |\lambda|^2 \omega.$$

So

$$\phi^* \tilde{\omega} = r \omega$$

for some r. Similarly

$$\Phi^* \widetilde{\Omega}_1 = \Phi^* \tilde{\lambda}(\tilde{\omega}_1 + \tilde{\mu}\tilde{\omega}) = \tilde{\lambda}(\phi^* \tilde{\omega}_1 + \tilde{\mu}\phi^* \tilde{\omega})$$

and

$$\Phi^*(\widetilde{\Omega}_1) = \Omega_1 = \lambda(\omega_1 + \mu\omega).$$

So

$$\phi^* \tilde{\omega}_1 = \lambda'(\omega_1 + \mu'\omega).$$

Thus ϕ is a CR map. It is a diffeomorphism since Φ is one.

EXERCISE 3. Show that Lemma 2 holds even if we only assume $\Phi^*(\widetilde{\Omega}_j) = \Omega_j$ for $j = 0$ and 1.

Now that we have related CR diffeomorphisms $\phi \colon M \to \widetilde{M}$ to maps $\Phi \colon B \to \widetilde{B}$ with $\Phi^*(\widetilde{\Omega}_j) = \Omega_j$, we can prove some interesting theorems. We start as usual with forms ω and ω_1 on M^3 satisfying $d\omega = i\omega_1\overline{\omega}_1$ (mod ω). On $M^3 \times \mathbb{C}^2 \times \mathbb{R}$ we have the well-defined forms $\Omega_0, \ldots, \Omega_4$. These give functions R, S, r as in the previous chapter. Recall that R is well-defined on B and that if we choose forms ω and ω_1 on M, which then lead to local coordinates on B, R takes the form

$$R = \frac{r(x)}{\overline{\lambda}^3 \lambda},$$

where $r(x)$ is a function on M. A different choice of ω and ω_1 would lead to a different function $\tilde{r}(x)$. That is, $r(x)$ is not an invariant at each x, but $r(x)$ and $\tilde{r}(x)$ are either both zero or both nonzero. So we call $r(x)$ a *relative invariant*. We now explore the consequences of having r either identically zero or nowhere zero on an open set. We start with a result analogous to Theorem 5.1.

THEOREM 1. *Any two CR manifolds with $r \equiv 0$ are locally equivalent.*

PROOF. Let $r \equiv 0$ on M. Thus $R \equiv 0$ on B and, according to an exercise from Chapter 6, also $S \equiv 0$. So the forms $\Omega_0, \ldots, \Omega_4$ satisfy

$$d\Omega_j = C^j_{k\ell}\Omega_k\Omega_\ell,$$

for some constants $C^j_{k\ell}$. If $\tilde{r} \equiv 0$ on $\widetilde{M}$ then we also have

$$d\tilde{\Omega}_j = C^j_{k\ell}\tilde{\Omega}_k\tilde{\Omega}_\ell,$$

with the same constants. By Exercise 5.3 we have a map $\Phi: B \to \tilde{B}$ with $\Phi^*(\tilde{\Omega}_j) = \Omega_j$. Thus by Lemma 3 there is a local CR diffeomorphism $\phi: M \to \widetilde{M}$ and we are done.

Since the hyperquadric Q has $r \equiv 0$ this theorem means that if $r \equiv 0$ on M then each point $p \in M$ has an open neighborhood U and a CR diffeomorphism $\phi: U \to V$ where V is an open set in Q. If, further, M is real analytic then ϕ is also real analytic (since it solves a real analytic Frobenius system) and thus by Theorem 1.1, ϕ extends to a holomorphic map. That is, if M is real analytic with $r \equiv 0$, then each $p \in M$ has an open neighborhood U' in $\mathbb{C}^2$ and a local biholomorphism $\psi: U' \to V'$ where V' is also an open set in $\mathbb{C}^2$ such that $\psi(M \cap U') = Q \cap V'$. It is necessary that M be real analytic in order to get the local biholomorphism:

EXERCISE 4. Show that there exists a C^∞ hypersurface M that is not real analytic but on which r is identically zero. (Hint: Construct M as an image of Q using a map, one component of which is a CR function that is not the restriction of a holomorphic function.) Note that since M is not real analytic, it cannot be the image, under a biholomorphism, of Q.

Note that if in the definition of Ω_2 we replace the variables λ and μ by known functions $\lambda(x)$ and $\mu(x)$ and the differential $d\lambda$ by $\frac{\partial\lambda}{\partial x_i}dx_i = \lambda_0\omega + \lambda_1\omega_1 + \lambda_{\bar{1}}\overline{\omega}_1$, then we obtain a one-form on M, and similarly for Ω_3 and Ω_4. We want to characterize the sets of one-forms on M that we can obtain in this way. We do not want to assume that M is realizable and so it is convenient to use the jet bundle formulation from Chapter 6 rather than rederive some of the Cartan formulas. Thus we start with forms ϕ and ϕ_1 with

$$d\phi = i\phi_1\overline{\phi}_1,$$
$$d\phi_1 = \alpha_{01}\phi\phi_1 + \alpha_{0\bar{1}}\phi\overline{\phi}_1 + \alpha_{1\bar{1}}\phi_1\overline{\phi}_1,$$

where these coefficients are considered as known quantities.

LEMMA 3. *Let the one-forms ϕ and ϕ_1 describe the given CR structure and, in addition, satisfy $d\phi = i\phi_1 \wedge \overline{\phi}_1$. There are unique forms ϕ_2, ϕ_3, ϕ_4 and unique functions $R(x)$ and $S(x)$ such that*

(1) ϕ_2 *is imaginary and ϕ_4 is real,*

(2) $d\phi_1 = -\phi_1\phi_2 - \phi\phi_3,$

(3) $d\phi_2 = 2i\phi_1\overline{\phi}_3 + i\overline{\phi}_1\phi_3 - \phi\phi_4,$

(4) $d\phi_3 = -\phi_1\phi_4 - \overline{\phi}_2\phi_3 - R\phi\overline{\phi}_1,$

(5) $d\phi_4 = i\phi_3\overline{\phi}_3 + (S\phi_1 + \overline{S}\,\overline{\phi}_1)\phi.$

PROOF. Let us consider uniqueness. Write

$$\phi_j = A_j\phi + B_j\phi_1 + C_j\overline{\phi}_1, \quad j = 2, 3, 4.$$

We have $\mathrm{Re}\, A_2 = 0$, $B_2 = -\overline{C}_2$, $\mathrm{Im}\, A_4 = 0$, $B_4 = \overline{C}_4$. We now do computations similar to those in Chapter 6. However, the present case is much simpler since we are working on M instead of on B. The uniqueness proof will be given as an exercise.

EXERCISE 5. (a) Substitute into (2) to show that C_2 and C_3 are determined and that $A_2 - B_3 = \alpha_{01}$.

(b) Next use (3) to show that A_2, A_3, B_3, and B_4 are determined.

(c) Use (4) to determine A_4 and R.

(d) Note that (5) then determines S.

Thus if the forms exist they are certainly unique. In fact, had we been a little more precise in these computations we would have also established the existence. However, it is simpler to argue as follows. Starting with ϕ and ϕ_1 construct the forms $\Omega = |\lambda|^2\phi$, etc., on B. Then pull these back to forms on M by setting $\lambda = 1$ and $\mu = 0$. Leave $\rho = \rho(x)$ arbitrary. Call the pull-back of Ω_j the form ϕ_j. Then equations (1) to (5) are satisfied for some functions R and S, except for the condition that ϕ_2 is imaginary. Now since $\lambda = 1$ and $\mu = 0$, equation (6.85) yields

$$\phi_2 = \overline{\alpha_{1\bar{1}}}\omega_1 - \alpha_{1\bar{1}}\overline{\omega}_1 + (\rho - \Gamma^0)\omega,$$

and so

$$\mathrm{Re}\,\phi_2 = \rho\omega.$$

So just set $\rho(x) = 0$. Thus (1)–(5) has a solution which is just the pull-back of $\Omega, \ldots, \Omega_4$ using $\lambda = 1$, $\mu = 0$, $\rho = 0$. And by uniqueness this is the only solution to these equations for a given ϕ and ϕ_1 with $d\phi = i\phi_1 \wedge \overline{\phi}_1$.

REMARK. Note that the equations

$$d\Omega = i\Omega_1\overline{\Omega}_1 - \Omega(\Omega_2 + \overline{\Omega}_2)$$

and

$$d\phi = i\phi_1\overline{\phi}_1$$

force $\mathrm{Re}\,\phi_2$ to be a multiple of ω. So the single condition $\rho = 0$ is enough to yield the two conditions in $\mathrm{Re}\,\phi_2 = 0$.

LEMMA 4. *If the relative invariant* $r \neq 0$ *then there are precisely two choices of* (ϕ, ϕ_1) *that satisfy*

(i) ϕ *and* ϕ_1 *describe the given CR structure*,

(ii) $d\phi = i\phi_1 \wedge \overline{\phi}_1$, *and*

(iii) *the unique function* R *above is identically one.*

PROOF. Start with some ω and ω_1 that describe the CR structure and satisfy $d\omega = i\omega_1\overline{\omega}_1$. Then we have $\Omega = |\lambda|^2\omega$, etc. As we have seen, the function R in (6.82) is given by $R(x, \lambda) = r(x)(\overline{\lambda}^3\lambda)^{-1}$. Let ε be a fixed choice of $+1$ or -1. Then

$$\lambda = \varepsilon(|r(x)|^{-1/2}\overline{r})^{1/2}$$

gives the two solutions to $R(x, \lambda(x)) \equiv 1$. Set

$$\phi = |\lambda(x)|^2 \omega \quad \text{and} \quad \phi_1 = \lambda(x)(\omega_1 + \mu(x)\omega),$$

with $\mu(x)$ to be determined. There are unique values of $\mu(x)$ and $\rho(x)$ that make $\Omega_2 + \overline{\Omega}_2 = 0$; namely,

$$\mu(x) = \frac{i}{4}\left(\frac{r_{\bar{1}}}{r} + \frac{\overline{r}_{\bar{1}}}{\overline{r}}\right),$$

$$\rho(x) = -\frac{1}{8}\left(\frac{r_0}{r} + \frac{\overline{r}_0}{\overline{r}}\right).$$

With these values for λ, μ, and ρ, set

$$\phi_j = \Omega_j, \quad j = 2, 3, 4.$$

Then ϕ_2, ϕ_3, ϕ_4 are the unique forms in Lemma 3 and $R(x)$ is identically one. The other choice for ε means starting with ϕ' and ϕ_1' (with $\phi' = \phi$ and $\phi_1' = -\phi_1$) and obtaining unique forms $\phi_2', \phi_3', \phi_4'$. Thus $\{\phi, \ldots, \phi_4\}$ and $\{\phi', \ldots, \phi_4'\}$ are the only sets of one-forms that satisfy (i), (ii), and (iii).

It is easy to see how these two sets are related. Note that $\rho(x)$ and $\mu(x)$ do not depend on ε. So

$$(6) \qquad \phi = \phi', \quad \phi_1 = -\phi_1', \quad \phi_2 = \phi_2', \quad \phi_3 = -\phi_3', \quad \phi_4 = \phi_4'.$$

Thus when $r \neq 0$ we can find precisely two sets of forms $(\phi, \ldots, \phi_4)$ and $(\phi', \ldots, \phi_4')$ that satisfy (1)–(5). Let us call these forms the *distinguished sets of one-forms*. When we do not explicitly need the other forms, we may just as well call $\{\phi, \phi_1\}$ and $\{\phi', \phi_1'\}$ the distinguished sets of one-forms. Let us emphasize that for the remainder of this chapter we work with CR manifolds with $r \neq 0$. Now let M and M' be such manifolds with $\{\omega, \omega_1\}$ a distinguished set of one-forms on M and $\{\theta, \theta_1\}$ a distinguished set of one-forms on M'.

LEMMA 5. *A map* $\phi: M \to M'$ *is a CR diffeomorphism if and only if it satisfies either*

$$\phi^*(\theta) = \omega \quad \text{and} \quad \phi^*(\theta_1) = \omega_1$$

or

$$\phi^*(\theta) = \omega \quad \text{and} \quad \phi^*(\theta_1) = -\omega_1.$$

PROOF. If ϕ satisfies either set of equations then it is a CR map. Further,

$$\phi^*(\theta \wedge \theta_1 \wedge \overline{\theta}_1) = \omega \wedge \omega_1 \wedge \overline{\omega}_1 \neq 0,$$

and so ϕ is a CR diffeomorphism. Conversely, let ϕ be any CR diffeomorphism. Consider the distinguished set $\{\theta, \ldots, \theta_4\}$. Define $\{\widetilde{\omega}, \ldots, \widetilde{\omega}_4\}$ by

$$\widetilde{\omega} = \phi^*(\theta), \quad \widetilde{\omega}_1 = \phi^*(\theta_1), \quad \ldots, \quad \widetilde{\omega}_4 = \phi^*(\theta_4).$$

Since (1)–(5) hold for $\{\theta, \ldots, \theta_4\}$ with $R \equiv 1$ and since ϕ is a diffeomorphism, these equations hold for $\{\tilde{\omega}_1, \ldots, \tilde{\omega}_4\}$. Further $d\tilde{\omega} = i\tilde{\omega}_1\overline{\tilde{\omega}}_1$. Since ϕ is also a CR map, $\tilde{\omega}$ and $\tilde{\omega}_1$ describe the CR structure of M. Thus by Lemma 4 we must have

$$\tilde{\omega} = \omega \quad \text{and} \quad \tilde{\omega}_1 = \pm\omega_1.$$

This proves our lemma.

REMARK. A similar normalization resulting in two possibilities occurs in [CM, p. 247] under the assumption that $F_{42}(0) \neq 0$. We shall see in the next chapter that r and F_{42} are essentially the same.

THEOREM 2. *Let $p \in M$ with $r(p) \neq 0$ and let M' be any other CR manifold. For each point $q \in M'$ there exist at most two CR diffeomorphisms of a neighborhood of p to a neighborhood of q that take p to q.*

COROLLARY. *Let $p \in M$ with $r(p) \neq 0$. There exists at most one CR automorphism of M, besides the identity, that leaves p fixed.*

REMARK. Of course, the theorem does not assert that given p and q there actually is a CR diffeomorphism taking p to q. In fact, a generic CR structure usually admits no local CR automorphisms except for the identity. We will see one explanation for this in a moment. Also, in Theorem 4 we consider what happens when the automorphism of the above corollary exists for each p.

PROOF OF THEOREM. Let $\{\omega, \omega_1\}$ and $\{\theta, \theta_1\}$ be distinguished sets and let $\phi\colon U \to V$ be a CR diffeomorphism of neighborhoods of p and q. Then, perhaps after replacing ω_1 by $-\omega_1$, we have

$$\phi^*(\theta) = \omega, \qquad \phi^*(\theta_1) = \omega_1, \qquad \phi^*(\overline{\theta}_1) = \overline{\omega}_1.$$

According to Exercise 5.3, this, together with $\phi(p) = q$, determines ϕ uniquely. Thus there are only the two possibilities for ϕ and the proof is done.

We will now look closer at our distinguished sets. Let us start with such a set $\{\omega, \ldots, \omega_4\}$. These are one-forms on the three-dimensional manifold M. Since $\omega \wedge \omega_1 \wedge \overline{\omega}_1 \neq 0$, each other form can be written as a linear combination $a_0\omega + a_1\omega_1 + a_{\overline{1}}\overline{\omega}_1$.

EXERCISE 6. Use (1), (2), (3) and $d\omega = i\omega_1\overline{\omega}_1$ to show

$$\omega_2 = \alpha\omega_1 - \overline{\alpha}\overline{\omega}_1 + i\beta\omega,$$
$$\omega_3 = i\gamma\omega_1 + \theta\overline{\omega}_1 + \eta\omega,$$
$$\omega_4 = -\frac{i}{2}\overline{\eta}\omega_1 + \frac{i}{2}\eta\overline{\omega}_1 + \zeta\omega,$$

with β, γ, and ζ real. Further, show that the other choice of a distinguished set gives

$$\beta' = \beta, \ \gamma' = \gamma, \ \theta' = \theta, \ \zeta' = \zeta \quad \text{and} \quad \alpha' = -\alpha, \ \eta' = -\eta.$$

The nine functions $\{\alpha, \overline{\alpha}, \beta, \gamma, \zeta, \theta, \overline{\theta}, \eta, \overline{\eta}\}$ are invariants (except possibly for sign) of a CR structure. (Recall that they are defined only when

$r \neq 0$.) If any three of these are independent, then these three provide an invariantly defined coordinate system (except possibly for sign). Any CR diffeomorphisms must be the identity in such coordinates. We use this to prove the following result.

THEOREM 3. *Let M and $\widetilde{M}$ be CR manifolds and let $V \subset \widetilde{M}$ be an open subset such that $\tilde{r}$ is different from zero everywhere on V. Further assume that three functions from $\{\tilde{\alpha}, \dots, \tilde{\tilde{\eta}}\}$ provide coordinates over V. For each point $p \in M$ there are at most two CR diffeomorphisms taking a neighborhood of p into V.*

PROOF. Let us start with an exercise. Note first that if a CR diffeomorphism exists then $r(p) \neq 0$.

EXERCISE 7. Show that if $\phi: U \to \phi(U) \subset \widetilde{U}$ is a CR diffeomorphism with $U \subset M$ and $\widetilde{U} \subset V \subset \widetilde{M}$ then

$$\beta = \tilde{\beta} \circ \phi, \quad \gamma = \tilde{\gamma} \circ \phi, \quad \theta = \tilde{\theta} \circ \phi, \quad \zeta = \tilde{\zeta} \circ \phi,$$

and

$$\alpha = \pm \tilde{\alpha} \circ \phi, \qquad \eta = \pm \tilde{\eta} \circ \phi.$$

Now use the correct three functions to introduce coordinates on $\widetilde{M}$. Say these functions are $\tilde{\alpha}, \tilde{\tilde{\alpha}}$, and $\tilde{\beta}$. Then

$$\alpha(p) = \pm \tilde{\alpha}(\phi(p)) \quad \text{and} \quad \beta(p) = \tilde{\beta}(\phi(p)),$$

and so $\phi(p)$ is determined as one of two possible points. Each of these points gives rise to a unique value for $\phi(x)$ for each x near p. This proves the theorem. Note that this procedure can always be used to construct some ϕ which can then be tested to see if it is a CR map.

REMARK. An automorphism ϕ must satisfy the nine equations

$$\alpha(\phi(x)) \;=\; \varepsilon\alpha(x),$$
$$\vdots$$
$$\zeta(\phi(x)) \;=\; \zeta(x).$$

So it is clear that for a "generic" CR structure, the only solution is $\phi(x) \equiv x$ (and $\varepsilon = +1$). This can be made rigorous by examining the particular form of the functions $\alpha(x), \dots, \zeta(x)$. In this way, one can also derive global results. See [BSW].

Now let us look at what might follow if we assume the two possible CR diffeomorphisms actually exist.

THEOREM 4. *Assume $r \neq 0$ on M. If for each $p \in M$ there is a local CR automorphism, different from the identity, that leaves p fixed, then M admits a transitive three-parameter group of local automorphisms.*

REMARK. This is the analogue of the standard result on the existence of isometric involutions. See, for example, [KN, p. 243].

PROOF. First we show that both α and η are identically zero. So we start with some CR diffeomorphism ϕ. Then

$$\phi^*(\omega) = \omega, \quad \phi^*(\omega_1) = \varepsilon\omega_1, \quad \alpha \circ \phi = \varepsilon\alpha, \quad \eta \circ \phi = \varepsilon\eta,$$

where ε is either $+1$ or -1. If $\varepsilon = +1$, then in the usual manner we may conclude from the first two equations that ϕ must be the identity. Thus if ϕ is not the identity but leaves some point p fixed, then $\varepsilon = -1$ and $\alpha(p) = \alpha(\phi(p)) = -\alpha(p)$ and $\eta(p) = \eta(\phi(p)) = -\eta(p)$. Thus $\alpha \equiv 0$ and $\eta \equiv 0$ on M.

LEMMA 6. If α and η are identically zero on some open set, then β, γ, θ, ζ are each a constant on this open set.

PROOF. We choose one of the two distinguished sets of one-forms. Then we have

$$d\omega = i\omega_1\overline{\omega}_1,$$
$$d\omega_1 = -\omega_1\omega_2 - \omega\omega_3,$$
$$d\omega_2 = 2i\omega_1\overline{\omega}_3 + i\overline{\omega}_1\omega_3 - \omega\omega_4,$$
$$d\omega_3 = -\omega_1\omega_4 - \overline{\omega}_2\omega_3 - \omega\overline{\omega}_1,$$
$$d\omega_4 = i\omega_3\overline{\omega}_3 - \omega(S\omega_1 + \overline{S}\overline{\omega}_1).$$

Also, since $\alpha = \eta = 0$, we have

$$\omega_2 = i\beta\omega,$$
$$\omega_3 = i\gamma\omega_1 + \theta\overline{\omega}_1,$$
$$\omega_4 = \zeta\omega.$$

These allow us, for instance, to compute $d\omega_2$ in two ways. What results is

(7) $$i(d\beta)\omega - \beta\omega_1\overline{\omega}_1 = 3\gamma\omega_1\overline{\omega}_1.$$

Let us again use the notation

$$df = f_0\omega + f_1\omega_1 + f_{\overline{1}}\overline{\omega}_1.$$

Then, as we see by equating coefficients in (7),

(8) $$\beta_1 = 0, \quad \beta_{\overline{1}} = 0, \quad \beta = -3\gamma.$$

The following exercise shows that from the first two equations, plus the fact that $\omega^\perp$ gives a nonintegrable distribution, we may conclude that β is a constant. Then of course γ is also a constant.

EXERCISE 8. Let ϕ_1, ϕ_2, ϕ_3 be one-forms on an open connected set U in $\mathbb{R}^3$ with $\phi_1 \wedge \phi_2 \wedge \phi_3$ nowhere zero and also $\phi_3 \wedge d\phi_3$ nowhere zero. For any function f write $df = \sum_1^3 f_i\phi_i$. Show that if $f_1 = 0$ and $f_2 = 0$ on U then $df = 0$ on U and hence f is a constant.

In the same way let us equate the two expressions we may find for $d\omega_3$. We obtain

$$\text{(9)} \qquad \theta_1 = 0, \qquad |\theta|^2 = \gamma^2 - \zeta, \qquad \theta_0 = 2i\beta\theta - 1.$$

Note that the first and last equations yield also that $\theta_{01} = 0$. Proceeding in the same way with $d\omega_4$ yields no new information.

EXERCISE 9. Compute $d(d\theta)$ and show that $\theta_{01} + \theta_{\bar{1}}\bar{\theta} = 0$.

Thus $\theta_{\bar{1}} = 0$, and since $\theta_1 = 0$, we have that θ is constant. Thus also ζ is constant. This proves Lemma 6.

It then follows that

$$\text{(10)} \qquad \begin{aligned} d\omega &= i\omega_1\bar{\omega}_1, \\ d\omega_1 &= c_1\omega\omega_1 - c_2\omega\bar{\omega}_1, \end{aligned}$$

where c_1 and c_2 are constants. Hence there is a transitive local three-dimensional group of diffeomorphisms preserving ω and ω_1. But any map preserving ω and ω_1 must be a CR map. This proves Theorem 4.

Let us show, in addition, that this group has two components. We have just now used that given any two points p and q there is a unique map ϕ such that $\phi(p) = q$, $\phi^*(\omega) = \omega$, and $\phi^*(\omega_1) = \omega_1$. (Recall that this is possible because the coefficients in (10) are constants.) Further, ϕ is a CR map. But we can also find some ψ such that $\psi(p) = q$, $\psi^*(\omega) = \omega$, and $\psi^*(\omega_1) = -\omega_1$. This map is also a CR map and so maps of this type provide another component of our group. There can be no other CR maps because the distinguished set $\{\omega, \omega_1\}$ must pull back to either itself or the distinguished set $\{\omega, -\omega_1\}$. Recall that this is all under the assumption that $r \neq 0$. It follows from the next exercise that if the group of local diffeomorphisms of some CR manifold has dimension larger than three, then its dimension must be eight.

EXERCISE 10. Assume that for each $p \in M$, M a nondegenerate CR manifold, there are two distinct CR diffeomorphisms leaving p fixed and assume that neither is the identity. Then M admits a transitive eight-parameter group of local diffeomorphisms. (Hint: Use Theorem 1.)

CHAPTER 8

Chains

In this chapter we show that the Cartan connection allows us to define an invariant family of curves, we identify these curves with the chains which appeared in the Moser normal form, and we explore properties and examples.

§1. The chains of Cartan and Moser. We naturally start with our basic system:

$$d\Omega = i\Omega_1\overline{\Omega}_1 - \Omega(\Omega_2 + \overline{\Omega}_2),$$
$$d\Omega_1 = -\Omega_1\Omega_2 - \Omega\Omega_3,$$
$$d\Omega_2 = 2i\Omega_1\overline{\Omega}_3 + i\overline{\Omega}_1\Omega_3 - \Omega\Omega_4,$$
$$d\Omega_3 = -\Omega_1\Omega_4 - \overline{\Omega}_2\Omega_3 - R\Omega\overline{\Omega}_1,$$
$$d\Omega_4 = i\Omega_3\overline{\Omega}_3 - (\Omega_2 + \overline{\Omega}_2)\Omega_4 + (S\Omega_1 + \overline{S}\,\overline{\Omega}_1)\Omega.$$

To write these forms in local coordinates we first choose (ω, ω_1) to give the CR structure with $d\omega = i\omega_1\overline{\omega}_1 \pmod{\omega}$. Then, using Cartan's formulation (see equation (6.50)),

$$\Omega = |\lambda|^2\omega,$$
$$\Omega_1 = \lambda(\omega_1 + \mu\omega),$$
$$\Omega_2 = \frac{d\lambda}{\lambda} + A\omega_1 + B\overline{\omega}_1 + C\omega,$$
$$\Omega_3 = \frac{1}{\overline{\lambda}}(d\mu + D\omega_1 + E\overline{\omega}_1 + F\omega),$$
$$\Omega_4 = \frac{1}{|\lambda|^2}\left(d\rho + G\omega + \frac{i\mu}{2}d\mu - \frac{i\overline{\mu}}{2}d\overline{\mu} + H\omega_1 + \overline{H}\,\overline{\omega}_1\right),$$

where the various coefficients depend on (x, μ, λ, ρ) and these variables are local coordinates for the eight-dimensional bundle B. Note that $\{\Omega_1, \Omega_3, \overline{\Omega}_1, \overline{\Omega}_3\}$ defines a closed differential ideal. We see from the above expressions for Ω_1 and Ω_3 that each four-dimensional integral submanifold N must contain the vectors $\{\frac{\partial}{\partial\lambda}, \frac{\partial}{\partial\overline{\lambda}}, \frac{\partial}{\partial\rho}\}$ in its tangent space at each of its points. So N must be of the form $\{(x(t), \mu(t), \lambda, \rho)\}$ with λ and ρ arbitrary. Note that $\omega(\frac{dx}{dt})$ cannot be zero and so $x(t)$ is a smooth curve in M and the natural projection $B \to M$ takes N onto this curve. Also note that for an arbitrary (x_0, μ_0) the three-dimensional surface $\{(x_0, \mu_0, \lambda, \rho) : \lambda$ and ρ are arbitrary$\}$ lies in some N.

DEFINITION. A *chain* is any curve in M obtained by the projection into M of an integral submanifold of the ideal $\{\Omega_1, \overline{\Omega}_1, \Omega_3, \overline{\Omega}_3\}$.

REMARKS. 1) We should be calling this a Cartan chain to distinguish it from the one defined in Chapter 4. And in turn this latter should be called a Moser chain. But, as we soon show, these two concepts actually are the same.

2) These chains generalize a concept in complex projective geometry (see the next chapter).

A first step towards understanding the chains is to realize that μ is a sort of slope, or, more precisely, a reciprocal slope. For, coordinates can be introduced so that, at a point,

$$\omega = du \quad \text{and} \quad \omega_1 = dz.$$

Thus for a curve $(z(t), u(t))$ the equation $\Omega_1 = 0$ defines μ at the given point to be

$$\mu = -\frac{\dot{z}}{\dot{u}},$$

and μ is the reciprocal of the usual complex slope. We have $\mu = 0$ for the u-axis and $|\mu| = \infty$ for directions in the z-plane. Any unoriented direction at a point x_0 not in the H plane has a finite slope μ_0 and so the unique integral manifold N through $(x_0, \mu_0, \lambda, \rho)$ projects onto a chain through x_0 and having direction μ_0. This chain is clearly unique. So we have the following result.

PROPOSITION 1. *Through each point $p \in M$ and each direction transverse to H there passes a unique chain.*

Note that chains have an obvious orientation obtained by requiring that $\omega(\frac{d\gamma}{dt}) > 0$, when ω is normalized so that $d\omega = i\omega_1\overline{\omega}_1$.

The chains live on M and can be defined directly as the solutions of certain equations. Throughout this chapter, we derive these equations in various degrees of explicitness. For the most part, we deal with embedded hypersurfaces and use the formulation of Cartan, especially the expressions (6.28), (6.29), etc., of the coefficients occurring in $\Omega, \Omega_1, \ldots, \Omega_4$. Indeed, we start by using these expressions to rewrite (6.50) as

(1)
$$\Omega_1 = \lambda(\omega_1 + \mu\omega),$$
$$\Omega_3 = \frac{1}{\lambda}(d\mu + \omega_3 + \mu\overline{\omega}_2 + (\rho + \frac{i}{2}|\mu|^2)\omega_1$$
$$+ i\mu^2\overline{\omega}_1 + \mu(\rho + \frac{i}{2}|\mu|^2)\omega),$$

where

(2)
$$\omega_2 = -b\omega_1 + \frac{1}{4}ic\omega,$$
$$\omega_3 = \frac{1}{4}ic\omega_1 + \frac{1}{6}\bar{\ell}\omega.$$

Recall that b, c, and ℓ are defined by

(3) $$d\omega = i\omega_1\bar{\omega}_1 + b\omega\omega_1 + \bar{b}\omega\bar{\omega}_1,$$

(4) $$db = b_0\omega + b_1\omega_1 + b_{\bar{1}}\bar{\omega}_1,$$

(5) $$c = b_{\bar{1}},$$

(6) $$\ell = c_1 - bc - 2ib_0,$$

and that c is real.

We seek a real curve on which Ω_1 and Ω_3 restrict to zero. Substituting $\omega_1 = -\mu\omega$ into (2), we obtain

$$\omega_2 = (\mu b + \frac{1}{4}ic)\omega,$$
$$\omega_3 = (-\frac{1}{4}ic\mu + \frac{1}{6}\bar{\ell})\omega,$$

and so

$$\Omega_1 = 0 \quad \text{and} \quad \Omega_3 = 0$$

become

(7)
$$\omega_1 = -\mu\omega,$$
$$d\mu = (i\mu|\mu|^2 - \bar{b}|\mu|^2 + \frac{1}{2}ic\mu - \frac{1}{6}\bar{\ell})\omega.$$

We have shown that the integral manifolds of $\{\Omega_1, \bar{\Omega}_1, \Omega_3, \bar{\Omega}_3\}$ project onto curves in M and that these curves satisfy (7). Let us change viewpoints and show directly that (7) defines a family of curves in M. We are thinking of ω and ω_1 as given forms and so b, c, and ℓ are considered given functions. We first seek a curve $\Gamma(t) = (\gamma(t), \mu(t)) \in M \times \mathbb{C}$ for which each equation in (7) holds when all the forms are evaluated on the vector $d\Gamma/dt$. We have already explained how μ can be thought of as a slope. At a point where $\omega = du$ and $\omega_1 = dz$ we have, from $\omega_1 = -\mu\omega$,

$$\mu = -\frac{\dot{z}}{\dot{u}}.$$

At nearby points we have

$$\mu(t) = f(\gamma(t), \dot{\gamma}(t)),$$

where

$$\gamma(t) = (z(t), u(t)).$$

So the second equation of (7) yields

(8) $$\frac{d}{dt}f(\gamma(t), \dot{\gamma}(t)) = F(\gamma(t), \dot{\gamma}(t)).$$

Hence $\gamma(t)$ is a chain if it satisfies a certain second-order equation.

Our next result has already been proved (in Moser's framework) at the end of Chapter 4.

PROPOSITION 2. *The chains on Q are the intersections of Q with complex lines.*

EXERCISE 1. Assume this result and show that the only unbounded chains on Q are the real lines $\{z$ fixed, u arbitrary$\}$. In particular, the u-axis is a chain.

REMARK. To facilitate comparisons with the Moser normal form, in this chapter we take as the hyperquadric

$$Q = \{(z_1, z_2) : \operatorname{Im} z_2 = |z_1|^2\}.$$

That is, we do not use the one-half factor.

PROOF. On the hyperquadric Q we may take $\omega = i\partial r = \frac{1}{2}dz_2 - i\bar{z}_1 dz_1$ and $\omega_1 = dz_1$. So $d\omega = i\omega_1\bar{\omega}_1$ and b is identically zero. Hence c and ℓ are also identically zero, and the equations for the chains are

$$\omega_1 = -\mu\omega,$$
$$d\mu = i\mu|\mu|^2\omega.$$

Now, on the complex line $\alpha z_1 + \beta z_2 = \gamma$, we have $\alpha dz_1 + \beta dz_2 = 0$, and so on the intersection of this line with Q, ω_1 and ω are related by

$$\alpha\omega_1 + \beta(2\omega + 2i\bar{z}_1\omega_1) = 0.$$

If we set $\mu = 2\beta/(\alpha + 2i\beta\bar{z}_1)$ we have

$$\omega_1 = -\mu\omega.$$

If we compute $d\mu$ and replace $d\bar{z}$ by $-\bar{\mu}\omega$ then we obtain

$$d\mu = i\mu|\mu|^2\omega.$$

Thus the intersection of the line with Q is a chain. We get all the chains this way since a direction transverse to H determines both a unique chain and a unique complex line.

REMARKS. 1) There is no analogue of Proposition 2 for a general hypersurface. We shall give an example of a chain on some M that is not the intersection of M with a Riemann surface. The example is very simple but relies on the fact that the Moser and Cartan chains coincide, so we defer it until later.

2) We note for later that $|\mu|$ is constant along any chain on the hyperquadric since

$$d|\mu|^2 = \bar{\mu}\,d\mu + \mu\,d\bar{\mu} = \bar{\mu}(i\mu|\mu|^2\omega) + \mu(-i\bar{\mu}|\mu|^2\omega) = 0.$$

It is easy to explicitly find all these chains. We restrict ourselves to chains through the origin. In terms of the coordinates (z, u) with $z = z_1$ and

$u = \operatorname{Re} z_2$, our forms are

$$\omega_1 = dz \quad \text{and} \quad \omega = \frac{1}{2}(du + iz\,d\bar{z} - i\bar{z}\,dz)$$

when we take Q to be $\{v = |z|^2\}$. It is convenient to choose a time parameter such that along a given chain we have

$$\omega = \frac{1}{|\mu|^2}, \quad \text{and so} \quad \omega_1 = -1/\bar{\mu} \quad \text{and} \quad d\mu = i\mu.$$

That is, we consider the system

$$\dot{u} = \frac{2}{|\mu|^2} - iz\,\dot{\bar{z}} + i\bar{z}\,\dot{z},$$

$$\dot{z} = -1/\bar{\mu},$$

$$\dot{\mu} = i\mu.$$

In doing this we lose the solution given by the u-axis since here $\mu = 0$. We add the initial conditions

$$u(0) = 0, \quad z(0) = 0, \quad \mu(0) = \nu.$$

The unique solution to this initial value problem is

$$z(t) = \frac{i}{\bar{\nu}}(e^{it} - 1), \quad u(t) = \frac{2 \sin t}{|\nu|^2}, \quad \mu(t) = \nu e^{it}.$$

Naturally, the equations for the chains on an arbitrary strictly pseudo-convex hypersurface are more complicated. As a first step towards writing these equations we express b, c, etc. in terms of a defining function. Our immediate goal is to show that the chains defined by Moser and those defined by Cartan are the same. So we start with a hypersurface in partial normal form

$$v = |z|^2 + \sum_{\substack{k \geq 2 \\ \ell \geq 2}} F_{k\ell}(u) z^k \bar{z}^\ell,$$

which we abbreviate as

(9) $$v(z, \bar{z}, u) = |z|^2 + F(z, \bar{z}, u).$$

At first we let F be more general than this partial normal form and only assume $F(z, \bar{z}, u) = O(|z|^K)$ for some $K \geq 4$. This will be useful later on.

We have seen in Chapter 1 that for any defining function r the form $i\partial r$ can be taken as ω as long as dz_2 is replaced by $d(u + iv(z, \bar{z}, u))$. Also, we can take dz_1 as ω_1. So we start with the defining function

(10) $$r = \frac{1}{2i}(z_2 - \bar{z}_2) - v(z, \bar{z}, u),$$

and compute that

$$i\partial r = \left(\frac{1}{2} + \frac{1}{2}v_u{}^2\right) du + \left(-\frac{i}{2} + \frac{1}{2}v_u\right) v_z dz$$
$$+ \left(\frac{i}{2} + \frac{1}{2}v_u\right) v_{\bar z} d\bar z.$$

(11)

We substitute (9) into (11) and use $F = O(|z|^K)$, $K \geq 4$. The result is

$$(12) \qquad i\partial r = \left(\frac{1}{2} + O(|z|^{2K})\right) du + \left(-\frac{i}{2}\bar z - \frac{i}{2}A + O(|z|^{2K-1})\right) dz$$
$$+ \left(\frac{i}{2}z + \frac{i}{2}\overline{A} + O(|z|^{2K-1})\right) d\bar z$$

with

$$(13) \qquad\qquad\qquad A = F_z + i\bar z F_u,$$

and so

$$(14) \qquad\qquad d(i\partial r) = i(1 + B)dz\,d\bar z$$
$$+ i\left(\frac{1}{2}A_u + O(|z|^{2K-1})\right)dz\,du,$$

with

$$B = \frac{1}{2}\left(A_{\bar z} + \overline{A}_z\right) + O(|z|^{2K-2}).$$

Since

$$du = (2 + O(|z|^{2K}))i\partial r + O(|z|)dz + O(|z|)d\bar z,$$

(14) may be rewritten as

$$d(i\partial r) = i(1 + B)dz\,d\bar z + (iA_u + O(|z|^K))dz(i\partial r),$$

where now B is of the form

$$(15) \qquad\qquad B = \frac{1}{2}\left(A_{\bar z} + \overline{A}_z\right) + O(|z|^K).$$

Now let $f = (1 + B)^{-1}$ and take

$$(16) \qquad\qquad\qquad \omega = if\,\partial r.$$

Thus

$$d\omega = idz\,d\bar z + (iA_u + O(|z|^K))dz\,\omega + \frac{df}{f}\omega.$$

Define the coefficients a and β by

$$\omega = a\,du + \beta\,dz + \overline{\beta}\,d\bar z.$$

From Exercise 6.3, we know that for any function h

$$dh = h_0\omega + h_1 dz + h_{\bar 1}d\bar z,$$

where

$$h_0 = a^{-1}h_u,$$
$$h_1 = h_z - a^{-1}\beta h_u,$$
$$h_{\bar{1}} = h_{\bar{z}} - a^{-1}\bar{\beta}h_u.$$

Since $f = 1 + O(|z|^{K-2})$ we have from (12) that

$$\omega = \left(\frac{1}{2} + O(|z|^{K-2})\right) du + \left(-\frac{i}{2}\bar{z} + O(|z|^{K-1})\right) dz$$
$$+ \left(\frac{i}{2}z + O(|z|^{K-1})\right) d\bar{z},$$

and so

$$a^{-1} = 2 + O(|z|^{K-2}),$$
$$a^{-1}\beta = -i\bar{z} + O(|z|^{K-1}),$$
$$a^{-1}\bar{\beta} = iz + O(|z|^{K-1}).$$

Thus

$$h_0 = (2 + O(|z|^{K-2}))h_u,$$
$$h_1 = h_z + (i\bar{z} + O(|z|^{K-1}))h_u,$$
$$h_{\bar{1}} = h_{\bar{z}} - (iz + O(|z|^{K-1}))h_u.$$

Hence

$$d\omega = i\,dz\,d\bar{z} + b\omega dz + \bar{b}\omega d\bar{z}$$

with

(17) $$b = -iA_u - f_z - Bf_z - i\bar{z}f_u + O(|z|^K).$$

This computation is valid for any hypersurface of the form (9) provided $F = O(|z|^K)$. The functions A, B, and f are defined in (13), (15), and (16). Further,

$$c = b_{\bar{1}} = b_{\bar{z}} - izb_u + O(|z|^{K-1}),$$
$$c_1 = c_z + i\bar{z}c_u + O(|z|^{K-1}),$$
$$c_{1\bar{1}} = c_{1\bar{z}} - izc_1 + O(|z|^{K-1}),$$
$$\ell = c_1 - bc - 2ib_0.$$

Since $B = O(|z|^{K-2})$ and $f = 1 - B + O(|B|^2)$, (17) can be rewritten, less precisely but more simply, as

$$b = -iA_u + B_z + O(|z|^{K-1}).$$

Corresponding to this simplification, we have

(17')
$$\begin{cases} b = F_{zz\bar{z}} + O(|z|^{K-1}), \\ c = F_{zz\bar{z}\bar{z}} + O(|z|^{K-2}), \\ c_1 = F_{zzz\bar{z}\bar{z}} + O(|z|^{K-3}), \\ c_{1\bar{1}} = F_{zzz\bar{z}\bar{z}\bar{z}} + O(|z|^{K-4}), \\ \ell = F_{zzz\bar{z}\bar{z}} + O(|z|^{K-3}). \end{cases}$$

Now let us assume that our hypersurface is in partial normal form in the sense that

(18)
$$F(z, \bar{z}, u) = \sum_{\substack{k \geq 2 \\ \ell \geq 2}} F_{k\ell}(u) z^k \bar{z}^\ell.$$

For F in normal form, $K = 4$. So in this case (17) determines b up to $O(|z|^4)$ and the coefficients from (18) that enter into (17) are those with $k + l \leq 6$. A straightforward computation yields

$$b = 4F_{22}\bar{z} + 12F_{32}|z|^2 + 6F_{23}\bar{z}^2$$
$$+ (18F_{33} - 16F_{22}^2 + 2iF_{22}')z\bar{z}^2 + 8F_{24}\bar{z}^3 + 24F_{42}z^2\bar{z} + O(|z|^4).$$

(Recall that $F_{k\ell}$ is a function of u, so F_{22}' denotes the first derivative.) We now easily derive

$$c = 4F_{22} + 12(F_{32}z + F_{23}\bar{z}) + 4(9F_{33} - 8F_{22}^2)|z|^2 + 24(F_{24}\bar{z}^2 + F_{42}z^2) + O(|z|^3),$$
$$l = 4(3F_{32} + 12F_{42}z + (9F_{33} - 3iF_{22}' - 4F_{22}^2)\bar{z}) + O(|z|^2).$$

For later use, we note that

$$c_0 = 8F_{22}' + 24(F_{32}'z + F_{23}'\bar{z}) + O(|z|^2),$$
$$c_{1\bar{1}} = 4(9F_{33} - 8F_{22}^2) + 4iF_{22}' + O(|z|).$$

The chains on the hypersurface

$$v = |z|^2 + F,$$

with F in partial normal form (18) therefore satisfy the equations (cf. (7)),

(19)
$$\begin{cases} dz = -\mu\omega, \\ d\mu = (-2F_{23}(u) + O(|z|) + O(|\mu|))\,\omega, \end{cases}$$

where $O(|\mu|)$ is small with respect to $|\mu|$.

It is now easy to see that the chains defined in Chapter 4 by mapping into normal forms are precisely the chains defined by Cartan. Let M^3 be any strictly pseudoconvex hypersurface in $\mathbb{C}^2$, p a point of M^3 and $v \in TM_p$ a direction transverse to H. There exists a unique curve γ through p and in the direction v that is a chain as defined in Chapter 4. Also, there is a unique curve Γ through p and in the direction v that is a chain as defined in this chapter. Let Φ be a local biholomorphism taking p to the origin, γ to the u-axis, and M to a hypersurface in normal form. The curve Γ maps to the unique curve in $\Phi(M)$ passing through the origin and tangent there to the u-axis, and which is a chain as defined in this chapter. So to show that $\gamma = \Gamma$ we need only show that the u-axis is a chain in the Cartan framework. But this is obvious since along the u-axis, $z \equiv 0$, and so also $\mu \equiv 0$; and $z \equiv 0$, $\mu \equiv 0$ is a solution of (19) as long as $F_{32}(u) \equiv 0$.

REMARK. The careful reader has noticed that the Moser chains were defined only for real analytic hypersurfaces whereas for the Cartan chains we need only C^∞ hypersurfaces, or even C^k hypersurfaces for k not too small. In fact, the Moser chains can also be introduced on C^k hypersurfaces by simply replacing M at each of its points by the hypersurface obtained by truncating any of its defining functions and then computing the second-order equation that defines the chain. But now we have an alternative to tracing through the constructions of Chapter 4 in order to see that this truncation method makes sense. For instead, we argue that the differential equations (8) giving the Cartan chains must also give the Moser chains in the C^k context.

This raises a natural question. When M is real analytic any chain γ can be mapped to the u-axis by a mapping taking M to some normal form. In particular, γ is the intersection of M with some Riemann surface $\mathscr{R}$ obtained in a geometrically significant way. Could this still be true for M of class C^∞? (Note that for M real analytic, γ is also real analytic and so for $\mathscr{R}$ we can take the complexification of γ. Viewed this way, i.e. without the geometric significance of the chain, the above question would not seem so natural!) In fact, a C^∞ chain does not necessarily lie on a Riemann surface or even on the boundary of a Riemann surface.

EXAMPLE. Let $F(u)$ be a C^∞ function with $F(u) = 0$ for $u \le 0$ and $F(u) > 0$ for $u > 0$. Let M be the hypersurface given by $v = |z|^2 + F(u)z^3\bar{z}^2$ and let γ be the chain through the origin and in the direction of the u-axis. Clearly γ coincides with the u-axis for $u < 0$. Assume there is some Riemann surface $\mathscr{R}$ such that γ is contained in $M \cap \mathscr{R}$. So $\mathscr{R}$ and $\{(0, w)\}$ have γ in common and so must coincide near γ and therefore in a full neighborhood of the origin. But $\{(0, w)\}$ intersects M in the u-axis and $\mathscr{R}$ intersects M in γ. Thus γ is the u-axis. But we have just seen that for the u-axis to be a chain we need $F_{32} \equiv 0$. Thus the u-axis cannot be a chain. This contradiction shows that γ cannot be contained in a Riemann surface. The same argument also shows that γ cannot be contained in the smooth boundary of an open Riemann surface.

We have seen that the Moser approach yields a projective structure on each chain. This projective structure is given by specifying the class of projective parametrizations. Now we show that the Cartan approach also yields a projective structure, but this time by specifying the forms Λ_1, Λ_2, Λ_3. Then we show that these two projective structures actually are the same.

(Caution, in [BS] it is remarked that this result is false in general, i.e., for higher-dimensional CR manifolds.)

Recall that the Cartan approach to chains realizes them as the projections of the integral manifolds of the closed differential system given by $\{\Omega_1, \overline{\Omega}_1, \Omega_3, \overline{\Omega}_3\}$. The equation for $d\Omega_2$ on such an integral manifold N is just

$$d\Omega_2 = -\Omega\Omega_4,$$

and so N is foliated by the three-dimensional integral manifolds of $\operatorname{Im}\Omega_2$. Each such manifold N^3 also projects to the underlying chain γ. On N^3 we have

(20)
$$\begin{cases} d\Omega = -2\Omega\operatorname{Re}\Omega_2, \\ d\operatorname{Re}\Omega_2 = -\Omega\Omega_4, \\ d\Omega_4 = -2(\operatorname{Re}\Omega_2)\Omega_4. \end{cases}$$

Set $\Lambda_1 = -\operatorname{Re}\Omega_2$, $\Lambda_2 = \frac{1}{\sqrt{2}}\Omega$, and $\Lambda_3 = -\sqrt{2}\Omega_4$. The equations (20) become

$$d\Lambda_1 = -\Lambda_2\Lambda_3,$$
$$d\Lambda_2 = -2\Lambda_1\Lambda_2,$$
$$d\Lambda_3 = 2\Lambda_1\Lambda_3.$$

Also,

$$\Lambda_2\Big|_{\pi^{-1}(q)} = 0,$$

for each $q \in \gamma$. Here π is the projection $N^3 \to M$.

Thus these forms define a projective structure on N^3 as defined in Chapter 5 and induce a projective parametrization on γ. We now want to explicitly determine this parametrization using the procedure outlined at the end of Chapter 5. Recall that in this method we start with a more or less arbitrary curve $\Gamma(\eta)$ in $\operatorname{PGL}(2)$ and try to determine its image $\sigma(\eta)$ in N^3 by solving ordinary differential equations. Then the map $\eta \to t(\eta)$ gives a map of $\mathbb{P}^1$ to the chain γ and is the sought projective parametrization.

Let us take for $\Gamma(\eta)$ the curve in $\operatorname{PGL}(2)$ given by $A = 1$, $D = 1$, $B = \eta$, $C = 0$. Thus on the tangent vector $\frac{d\Gamma}{d\eta}$ we have

$$\Lambda_{11}\left(\frac{d\Gamma}{d\eta}\right) = 0, \qquad \Lambda_{12}\left(\frac{d\Gamma}{d\eta}\right) = 1, \qquad \Lambda_{21}\left(\frac{d\Gamma}{d\eta}\right) = 0,$$

where $\Lambda_{11} = D\,dA - B\,dC$, etc. See Chapter 5. We seek a curve $\sigma(\eta)$ in N^3 with

(21) $\qquad \Lambda_1\left(\dfrac{d\sigma}{d\eta}\right) = 0, \qquad \Lambda_2\left(\dfrac{d\sigma}{d\eta}\right) = 1, \qquad \Lambda_3\left(\dfrac{d\sigma}{d\eta}\right) = 0.$

Note that in N^3, i.e., above the chain $\gamma(t)$, we have that $z(t)$, $u(t)$, $\mu(t)$ are functions determined by the equations

(22) $\qquad\qquad \omega_1 = -\mu\omega, \qquad d\mu = (i\mu|\mu|^2 + R)\omega$

with

$$R = -\bar{b}|\mu|^2 + \frac{1}{2}ic\mu - \frac{1}{6}\bar{\ell},$$

where b, c, and ℓ are given in (3)–(6). We substitute (22) into the expressions for Ω, Ω_2, and Ω_4 given in (6.50) to obtain

$$\Omega = |\lambda|^2\omega,$$

$$\Omega_2 + \overline{\Omega}_2 = d(\ln|\lambda|^2) + (2\rho + S)\omega,$$

(23)
$$|\lambda|^2\Omega_4 = d\rho + \left\{ T_1 + T_2\mu + \overline{T}_2\bar{\mu} + \rho^2 + \rho S \right.$$

$$\left. + \left(\frac{i}{2}(b\mu - \bar{b}\bar{\mu}) - \frac{3c}{4} \right) |\mu|^2 + \frac{1}{4}|\mu|^4 \right\}\omega,$$

where

(24)
$$\begin{cases} T_1 = \dfrac{11}{48}c^2 + \dfrac{1}{6}\left(|b|^2c + b\bar{\ell} + \bar{b}\ell - c_{1\bar{1}} + \dfrac{1}{2}ic_0 \right), \\[2mm] T_2 = \dfrac{i}{2}\overline{R} + i\dfrac{(\ell + 4ib_0)}{12} + \dfrac{1}{6}i\ell, \\[2mm] S = b\mu + \bar{b}\bar{\mu}. \end{cases}$$

In particular, if $b \equiv 0$ then R, S, T_1, and T_2 are all zero and the equations become

$$\Omega = |\lambda|^2\omega,$$

$$\Omega_2 + \overline{\Omega}_2 = d(\ln|\lambda|^2) + 2\rho\omega,$$

$$|\lambda|^2\Omega_4 = d\rho + \left\{ \rho^2 + \frac{1}{4}|\mu|^4 \right\}\omega,$$

while if $\mu = 0$ the equations become

$$\Omega = |\lambda|^2\omega,$$

$$\Omega_2 + \overline{\Omega}_2 = d(\ln|\lambda|^2) + 2\rho\omega,$$

$$|\lambda|^2\Omega_4 = d\rho + \{T_1 + \rho^2\}\omega.$$

We choose local coordinates on N^3. Let $\gamma(t)$ be the underlying chain; so μ, b, etc. are functions of t. Note that only $|\lambda|^2$ appears in our forms; so let $r = |\lambda|^2$ be the second coordinate, and ρ the third, i.e., $\gamma(t) = (t, r(t), \rho(t))$. From (21) and (23) we see that we are looking for a curve $\sigma(\eta) = (t(\eta), r(\eta), \rho(\eta))$ with

$$r\omega\left(\frac{d\sigma}{d\eta}\right) = \sqrt{2},$$

$$(d(\ln r) + (2\rho + S)\omega)\left(\frac{d\sigma}{d\eta}\right) = 0,$$

$$\left(d\rho + \left(\rho^2 + \rho S + \frac{1}{4}|\mu|^4 + T \right)\omega \right)\left(\frac{d\sigma}{d\eta}\right) = 0,$$

where

$$(24')\qquad T = T_1 + T_2\mu + \overline{T}_2\overline{\mu} + \left(\frac{i}{2}(b\mu - \overline{b}\overline{\mu}) - \frac{3c}{4}\right)|\mu|^2.$$

It is natural to at first choose the parametrization of the chain by requiring $\omega(\frac{d}{dt}) = 1$. Also, it will be convenient to determine η as a function of t rather than t as a function of η. Thus

$$\frac{d\sigma}{d\eta} = \left(\frac{d\eta}{dt}\right)^{-1}\frac{d\sigma}{dt}$$

and

$$\omega\left(\frac{d\sigma}{d\eta}\right) = \left(\frac{d\eta}{dt}\right)^{-1}.$$

Now we are looking for a curve $\sigma(t) = (t, r(t), \rho(t))$ and a function $\eta(t)$ satisfying (primes denote t-derivatives)

$$(25)\qquad \begin{cases} r = \sqrt{2}\,\eta', \\[2mm] \dfrac{r'}{r} + (2\rho + S) = 0, \\[2mm] \rho' + \rho^2 + \rho S + \dfrac{1}{4}|\mu|^4 + T = 0. \end{cases}$$

It is easy to eliminate r and ρ. We end up with an equation for $\eta(t)$,

$$(26)\qquad \mathscr{S}(\eta) = \frac{1}{4}|\mu|^4 + T - \frac{1}{4}S^2 - \frac{1}{2}S',$$

where $\mathscr{S}$ is the Schwarzian,

$$(27)\qquad \mathscr{S}(\eta) = \frac{1}{2}\left(\frac{\eta'''}{\eta'}\right) - \frac{3}{4}\frac{\eta''^2}{\eta'^2}.$$

Note that T and S are identically zero if $b \equiv 0$.

The solutions to (26) provide the parametrizations of γ. Any two solutions differ by a projective transformation, so the totality of solutions does provide us with a projective parametrization of γ. Equation (26) has also been derived, in a slightly different manner, by Kruzhilin [Kr, p. 537].

EXERCISE 2. If $\eta(t)$ and $\xi(t)$ are solutions to (26) for some functions μ, S, and T, show that there exist constants a, b, c, and d with

$$\eta = \frac{a\xi + b}{c\xi + d}.$$

(Hint: This is a general property of the Schwarzian. See also Exercise 4.6.)

Let us summarize. On N^3 we have a projective structure. If we choose a point $p \in \mathrm{PGL}(2)$ and a point $q \in N^3$, then there is a unique map $\psi: \mathrm{PGL}(2) \to N^3$ that takes p to q and preserves the forms. If $\tilde{\psi}$ comes from another choice of p and q then $\tilde{\psi} = \psi \circ g$ where $g \in \mathrm{PGL}(2)$. The map ψ passes to the quotient to give a map $\psi_1: \mathbb{P}^1 \to \gamma$ of the projective line to the chain. Again we have $\tilde{\psi}_1 = \psi_1 \circ g$ where now g acts on $\mathbb{P}^1$. In this way we have a well-defined projective parametrization of the chain. To actually see what this parametrization is, we choose any parametrization of the chain γ and we choose a curve $\Gamma(\eta)$ in $\mathrm{PGL}(2)$. We end up with (26). These choices allow us to determine $\eta\colon \mathbb{P}^1 \to \gamma$. The projective parametrization of γ given by $\{\eta \circ g \mid g \in \mathrm{PGL}(2)\}$ is well-defined and depends on neither of these choices.

We now show that this projective parametrization coincides with the one found in Chapter 4. So let M be a strictly pseudoconvex hypersurface, p a point of M, and γ a chain through p. Without loss of generality, we may take M to be in normal form with $p = 0$ and γ given by the u-axis. The projective parametrization introduced by Moser is represented by the usual parametrization of the u-axis. So we need show that when M is in normal form the projective parametrization, in the sense of Cartan, of the u-axis is also given by the usual parametrization. At first we work with

$$v = |z|^2 + \sum_{\substack{k \geq 2 \\ \ell \geq 2}} F_{k\ell} z^k \overline{z}^\ell,$$

without assuming $F_{22}(u) = F_{32}(u) = F_{33}(u) = 0$. Recall that the u-axis is a chain provided $F_{32}(u) = 0$, cf. Equation (19). Since

$$\omega = (1 + B)^{-1}(i\partial r),$$

we see that along the u-axis

$$\omega = \frac{1}{2}du,$$

and so we parametrize the u-axis by $u(t) = 2t$ in order to have $\omega(\frac{d}{dt}) = 1$. Of course if this parametrization is projective in the sense of Cartan, then so is the usual parametrization $u(t) = t$. The u-axis is given by $\mu = 0$. So we need to show that the function $\eta(t) = t$ satisfies (26) when μ is set equal to zero. Because of (24) and (24'), we need only show that T_1 is zero.

We have determined b, c, c_0, and $c_{1\overline{1}}$ immediately after (18). For convenience we rewrite these equations as

$$b = O(|z|),$$
$$c = 4F_{22} + O(|z|),$$
$$c_0 = 8F'_{22} + O(|z|),$$
$$c_{1\overline{1}} = 4(9F_{33} - 8F_{22}^2) + 4iF'_{22} + O(|z|).$$

Clearly $T_1 = 0$ provided $z = 0$, $\mu = 0$, $F_{22}(u) = 0$, and $F_{33}(u) = 0$.

As our first example of projective parametrizations, let us determine this parametrization for the chains on the hyperquadric Q. We shall see that in this case there is a simple geometric interpretation. We start with the usual choices,

$$\omega = \frac{1}{2}(du + iz d\bar{z} - i\bar{z} dz),$$

$$\omega_1 = dz.$$

Thus

$$d\omega = i\omega_1 \bar{\omega}_1$$

and b is identically zero. Hence (26) becomes (see the remark after (27))

(28) $$\mathscr{S}(\eta) = \frac{1}{4}|\mu|^4.$$

We seek $\eta = \eta(t)$ where t is the chain parametrization chosen to satisfy $\omega = 1$. We know that $|\mu|$ is a constant along each chain in Q. So it is easily seen that $\eta(t) = e^{|\mu|^2 it}$ is a solution to (28) as is any fractional linear transformation of $\eta(t)$. But we need that the solution be real. For instance,

(29) $$\eta(t) = i\frac{(1 - e^{i|\mu|^2 t})}{(1 + e^{i|\mu|^2 t})} = \frac{\sin(|\mu|^2 t)}{1 + \cos(|\mu|^2 t)}$$

is one real solution to (28). The real projective functions $\frac{a\eta + b}{c\eta + d}$ provide all other real solutions of (28) and give the projective parametrization of the chain.

Earlier we have used as our representation of the chains

$$z(t) = \frac{i}{\bar{\nu}}(e^{it} - 1), \qquad u(t) = \frac{2 \sin t}{|\nu|^2}, \qquad \mu(t) = \nu e^{it},$$

with t chosen so that $\omega = \frac{1}{|\mu|^2}$. But now $\omega = 1$. So we take as our chains through the origin

(30)
$$z(t) = \frac{i}{\bar{\nu}}(e^{i|\nu|^2 t} - 1),$$

$$u(t) = \frac{2 \sin (|\nu|^2 t)}{|\nu|^2}$$

(we can neglect the equation for $\mu(t)$).

Any chain is the intersection of Q with a complex line. We claim that each such chain is a circle (or a real line) in the complex line. For instance (30), to which we will confine our attention, is given by the circle

(31) $$\left|\zeta + \frac{i}{|\nu|^2}\right|^2 = \frac{1}{|\nu|^4},$$

where ζ is the coordinate on $2z_1 + \nu z_2 = 0$ given by $\zeta \to (\nu\zeta, -2\zeta)$. (This circle passes through the origin.)

Of course, a circle in $\mathbb{C}$ admits a projective parametrization of geometric significance, namely the one given by biholomorphisms taking the real axis to the circle. One such map for (31) is

(32) $$\zeta_0(\xi) = -\frac{2i}{|\nu|^2}\left(\frac{\xi}{\xi + i}\right), \qquad \xi \in \mathbb{R},$$

and any other such map is given by

$$\zeta(\xi) = \zeta_0 \left(\frac{a\xi + b}{c\xi + d} \right),$$

with a, b, c, d all real. When the circle is parametrized as in (30) then from $\nu\zeta = z$ we have

$$\zeta_0(\xi) = \frac{i}{|\nu|^2}(e^{i|\nu|^2 t} - 1).$$

Combining this with (32), we obtain

$$\xi = \frac{i(1 - e^{i|\nu|^2 t})}{(1 + e^{i|\nu|^2 t})}.$$

Thus the particular parametrization $\zeta_0(\xi)$ of the circle coincides with the particular representative (29) of the projective parametrization as a chain. And further, each parametrization of the circle using the biholomorphic image of the real line coincides with some solution of (28).

§2. **Example: A family of Reinhardt hypersurfaces.** The chains are relatively simple geometric objects that are invariant under CR diffeomorphisms. It is thus natural to use them to study when two given hypersurfaces are CR equivalent. As an example, we find a family of hypersurfaces that are all locally equivalent but no two of which are globally equivalent. We do this by computing the chains and their projective parametrizations for certain Reinhardt hypersurfaces.

We have seen in Chapter 1 that for the Reinhardt hypersurface

$$\mathscr{R}_B : |z_2| = |z_1|^B e^{|z_1|^2}, \quad z_1 \neq 0,$$

where B is a constant, the CR structure is given by forms

$$(33) \qquad \omega = -\frac{1}{2}(d\theta_2 - (B + 2e^{2s})d\theta_1),$$

$$(34) \qquad \omega_1 = dz_1 = e^{s+i\theta_1}(ds + id\theta_1),$$

and these forms satisfy

$$(35) \qquad d\omega = i\omega_1 \overline{\omega}_1.$$

We note first that if $B_1 \equiv B_2 \pmod{Z}$, i.e., if $B_1 - B_2$ is an integer, then

$$z_1 = z, \quad z_2 = z^{B_2 - B_1} w$$

is an automorphism of $\mathbb{C}^2 - \{(z, 0)\}$ and a CR diffeomorphism of $\mathscr{R}_{B_1}$ and $\mathscr{R}_{B_2}$. In particular, when B is an integer, $\mathscr{R}_B$ is globally equivalent to

$$\mathscr{R}_0 : |w| = e^{|z|^2}, \quad z \neq 0.$$

Further, there is a CR covering map of $Q - \{(0, u) : u \text{ real}\}$ onto each $\mathscr{R}_B$, B an integer, given by

$$(36) \qquad z_1 = z, \quad z_2 = z^B e^{-iw}.$$

Finally, note that $\mathscr{R}_0$ extends above $z = 0$ to give an everywhere smooth hypersurface $(z, e^{|z|^2 + i\theta})$ which is covered by Q.

LEMMA 1. *Each $\mathscr{R}_B$ is locally CR diffeomorphic to the hyperquadric Q and thus $\mathscr{R}_{B_1}$ and $\mathscr{R}_{B_2}$ are always locally CR equivalent.*

PROOF. This is easy since (36) is a local diffeomorphism even for B not an integer. Or, from (35) we see that $b \equiv 0$. Thus R and S in the structure equations (6.51) are also identically zero, and $\mathscr{R}_B$ is locally equivalent to Q.

A detailed study of the chains on $\mathscr{R}_B$ will allow us to prove the following result.

THEOREM 1. *$\mathscr{R}_{B_1}$ and $\mathscr{R}_{B_2}$ are globally CR equivalent if and only if $B_1 \equiv B_2 \pmod{Z}$.*

REMARK. We actually show that if $\mathscr{R}_{B_1}$ and $\mathscr{R}_{B_2}$ are CR equivalent on a large enough set then $B_1 \equiv B_2 \pmod{Z}$.

A simpler study of the chains, which we do first as a warm up, serves to prove the next result. Let T_s denote the torus in R_B that is defined by

$$z_1 = e^{s + i\theta_1}, \qquad z_2 = |z_1|^B e^{|z_1|^2} e^{i\theta_2},$$

with s held constant.

We shall see that T_s provides an obstruction to enlarging a local CR equivalence between open sets of $\mathscr{R}_B$ and Q.

THEOREM 2. *Let U be an open set in some $\mathscr{R}_B$. If U is CR diffeomorphic to an open set in Q then U cannot contain any torus T_s.*

One special case of Theorem 2 can be handled directly without recourse to chains.

PROOF OF THEOREM 2 (FOR B AN INTEGER). We take B to be zero and start with the covering map $\Phi : Q \to \mathscr{R}_0$ given by

$$z_1 = z, \qquad z_2 = e^{-iw}.$$

On any T_s we have

$$z_1 = e^s e^{i\theta_1} \quad \text{and} \quad z_2 = e^{|z_1|^2} e^{i\theta_2},$$

and so

$$\Phi^{-1}(T_s) = \{(z, w) : z = e^s e^{i\theta_1}, \ w = -\theta_2 + i|e^s|^2 + 2\pi N\}.$$

There clearly is no closed set in $\Phi^{-1}(T_s)$ that is mapped bijectively to T_s. Thus there is no open set $V \subset Q$ that Φ maps diffeomorphically onto a neighborhood of T_s. We now want to show a similar result holds for any CR diffeomorphism $\Psi : V \to \mathscr{R}_0$ and any open set $V \subset Q$. So we assume that $\Psi(V)$ contains some T_s and we derive a contradiction. Fix some $p_1 \in V$, $p_2 \in Q$ such that $\Psi(p_1) = \Phi(p_2)$. Let $\tilde{V}_2$ be a sufficiently small neighborhood of p_2 so that $\Phi \mid_{\tilde{V}_2}$ is a diffeomorphism. Call this diffeomorphism $\tilde{\Phi}$. Define $\tilde{V}_1$ by $\Psi(\tilde{V}_1) = \Phi(\tilde{V}_2)$ and set $\tilde{\Psi} = \Psi \mid_{\tilde{V}_1}$. So $\tilde{\Phi}^{-1} \circ \tilde{\Psi}$ is a CR diffeomorphism of $\tilde{V}_1$ to $\tilde{V}_2$. Hence there is a unique $g \in \mathrm{Aut}(Q)$, $g(\tilde{V}_1) = \tilde{V}_2$, such that $\tilde{\Psi} = \tilde{\Phi} \circ g$. It is easy to see that, as long as V is connected, $\Psi = \Phi \circ g$ on all of V. Thus if $\Psi(V)$ were to contain some T_s, then Φ would map some closed subset of $g(V)$ bijectively onto T_s. But we have seen that this is impossible.

So now we consider the case where B is not an integer. Here we begin our study of chains on $\mathscr{R}_B$. We again start with the forms

(37)
$$\omega = -\frac{1}{2}(d\theta_2 - (B + 2e^{2s})d\theta_1),$$
$$\omega_1 = e^{s+i\theta_1}(ds + id\theta_1),$$

which satisfy

$$d\omega = i\omega_1\overline{\omega}_1 \quad \text{and} \quad d\omega_1 = 0.$$

Because $b \equiv 0$, the chain equations are simply

(38)
$$\omega_1 = -\mu\omega, \quad d\mu = i\mu|\mu|^2\omega.$$

For our proof of the theorem we need only consider the chains that remain on the torus T_s. For these chains, (37) becomes

(39)
$$\omega = -\frac{1}{2}(d\theta_2 - (B + 2e^{2s})d\theta_1),$$
$$\omega_1 = e^{s+i\theta_1}(id\theta_1) = izd\theta_1,$$

and (38) becomes

(40)
$$izd\theta_1 = -\mu\omega, \quad d\mu = i\mu|\mu|^2\omega.$$

It is easy to see from this that the torus T_s is foliated by each of two families of chains:
   a) $\theta_1 = C$,
   b) $\theta_2 = B\theta_1 + C$,
where C is an arbitrary real constant. Consider the two chains we get by taking $C = 0$ in both a) and b). If B is not an integer, then the chain $\gamma = \{\theta_1 = 0,\ \theta_2 = t\}$ intersects the chain $\Gamma = \{\theta_1 = t,\ \theta_2 = Bt\}$ in at least two distinct points. (Here of course we identify θ_j with $\theta_j + 2\pi N$.) Now we give the proof of Theorem 2. If $\Phi: U \to Q$ is a CR diffeomorphism of U onto some open set of Q and if $T_s \subset U$ then $\Phi(\gamma)$ and $\Phi(\Gamma)$ are two different chains in Q that meet in two distinct points. This is impossible since each chain in Q is determined by a complex line which in turn is determined by any two of its points. Thus if $\Phi(\gamma)$ and $\Phi(\Gamma)$ had two points in common it would follow that they were identical.

EXERCISE 3. Show that the chains in these two families are the only chains that remain on T_s.

To prove Theorem 1 we consider all chains and not just ones restricted to lie on T_s. We also make use of projective parameters.

The equations of the chains are

$$\omega_1 = -\mu\omega, \qquad d\mu = i\mu|\mu|^2\omega.$$

That is, using our usual parametrization $\omega\left(\frac{d}{dt}\right) = 1$,

(41)
$$\frac{dz}{dt} = -\mu, \qquad \frac{d\mu}{dt} = i\mu|\mu|^2,$$

and so

(42)
$$\mu(t) = \nu e^{it|\nu|^2}, \qquad \nu = \text{constant}.$$

If $\nu = 0$, then z is a constant and we get one of the previously considered chains:

$$z = z_0, \qquad w = w_0 e^{i\theta_2}.$$

If $\nu \neq 0$, then

(43)
$$z(t) = z_0 + \frac{i}{\nu}\left(e^{it|\nu|^2} - 1\right)$$

and

$$w(t) = |z|^B e^{|z|^2} e^{i\theta_2(t)}$$
$$= e^{sB} e^{e^{2s}} e^{i\theta_2},$$

since $z = e^{s+i\theta_1}$. It is now easy to see that

$$\frac{dw}{w} = (B + 2e^{2s})ds + id\theta_2.$$

Recall that

$$\omega = -\frac{1}{2}(d\theta_2 - (B + 2e^{2s})d\theta_1),$$

and so

$$i\omega + \frac{1}{2}\frac{dw}{w} = \frac{1}{2}(B + 2e^{2s})\frac{dz}{z}.$$

Since $\omega\left(\frac{d}{dt}\right) = 1$, this becomes

$$\frac{d}{dt}\ln w = -2i + (B + 2e^{2s})\frac{d}{dt}\ln z.$$

Since s is a function of t, we cannot immediately integrate. But note that

$$e^{2s}\frac{d}{dt}\ln z = |z|^2\frac{\dot{z}}{z} = \bar{z}\,\dot{z}$$
$$= -(\bar{z}_0\nu + i)e^{it|\nu|^2} + i,$$

and so (44) becomes

$$\frac{d}{dt}\ln w = B\frac{d}{dt}\ln z - 2(\bar{z}_0\nu + i)e^{it|\nu|^2}.$$

We integrate this and use (43) to obtain

(44) $$w(t) = cz^B e^{2(\bar{z}_0 + \frac{i}{\nu})z}.$$

At $t = 0$, $|w| = |z_0|^B e^{|z_0|^2}$. This determines $|c|$ and yields

(45) $$w(t) = z^B e^{2(\bar{z}_0 + \frac{i}{\nu})(z - z_0)} e^{i\sigma} e^{|z_0|^2}.$$

For arbitrary real σ and complex ν, $\nu \neq 0$, this gives the chains on $\mathscr{R}_B$.

There is an interesting failure of continuity for these solutions that will soon be important to us.

EXAMPLE. The solution of (41) with the initial conditions

$$z(0) = 0, \quad \mu(0) = \nu$$

is

$$z(t, \nu) = \frac{i}{\nu}(e^{it|\nu|^2} - 1), \quad \mu(t, \nu) = \nu e^{it|\nu|^2},$$

except that for $\nu = 0$ the solution is

$$z(t, 0) = 0, \quad \mu(t, 0) = 0.$$

Let Γ_ν be the subset of $\mathbb{C}$:

$$\Gamma_\nu = \{z(t, \nu): -\infty < t < \infty\}.$$

Then

$$\lim_{\nu \to 0} \Gamma_\nu \neq \Gamma_0.$$

We now derive the projective parametrizations of the chains. As we have seen in (26), if a chain is parametrized so that $\omega\left(\frac{dy}{dt}\right) = 1$, then the projective parametrizations are the solutions of

$$\mathscr{S}(\eta(t)) = \frac{1}{4}|\mu|^4 + U,$$

where again $\mathscr{S}(\eta)$ is the Schwarzian

$$\mathscr{S}(\eta) = \left(\frac{\eta'''}{2\eta'}\right) - \frac{3}{4}\frac{\eta''^2}{\eta'^2},$$

and $U \equiv 0$ if $b \equiv 0$. So for the class of Reinhardt domains we are considering

(46) $$\mathscr{S}(\eta) = \frac{1}{4}|\mu|^4.$$

We already know that $|\mu|$ is a constant (see (42)). If this constant is zero, then the solutions to (46) are

$$\eta(t) = \frac{at + b}{ct + d},$$

for real constants a, b, c, d. So $\eta: \gamma \to \mathbb{P}^1$ covers $\mathbb{P}^1$ once as t varies from $-\infty$ to $+\infty$. In particular, t itself is a projective parameter.

Recall that the chain with $\mu = 0$ is

$$(47) \qquad\qquad z = z_0, \qquad w = w_0 e^{i\theta_2}.$$

The dependence of θ_2 on t is easily determined: From $\omega\left(\frac{d\gamma}{dt}\right) = 1$ and $d\theta_1 = 0$ we obtain $\frac{d\theta_2}{dt} = -2$ and so $\theta_2 = -2t$. Notice that this chain has period π in t and so the chain is covered infinitely often as t varies from $-\infty$ to $+\infty$.

If $\mu \neq 0$ then one solution to (46) is

$$(48) \qquad\qquad \eta(t) = \frac{\operatorname{Im} e^{i|\mu|^2 t}}{1 + \operatorname{Re} e^{i|\mu|^2 t}} = \frac{\sin(|\mu|^2 t)}{1 + \cos(|\mu|^2 t)},$$

and all other solutions are of the form

$$\frac{a\eta + b}{c\eta + d},$$

with a, b, c, and d being real constants. In particular, η goes from $-\infty$ to $+\infty$ as t goes from $\frac{-\pi}{|\mu|^2}$ to $\frac{+\pi}{|\mu|^2}$. Thinking of $\eta(t)$ as taking values in $\mathbb{P}^1$, we see that η covers $\mathbb{P}^1$ once as t changes by $\frac{2\pi}{|\mu|^2}$.

From (43) we know that as t changes by $\frac{2\pi}{|\mu|^2}$, $z(t)$ returns to its original value. Thus η covers $\mathbb{P}^1$ once as $z(t)$ varies along the closed curve.

We now distinguish four types of chains. We shall soon see that a slight modification of this classification is a true CR construction in that it uses only CR data and, in particular, does not use projection onto the z-plane. The four classes of chains are:

1. The chain is given by $z \equiv z_0$, θ_2 arbitrary.

2. The chain when projected into the z-plane is a simple closed curve that does not contain the origin in its interior.

3. The chain when projected into the z-plane is a simple closed curve that contains the origin in its interior.

4. The chain when projected into the z-plane approaches the origin.

These classes are clearly mutually exclusive. And it is clear from (43), which is valid for all chains except those of class one, that these classes are also exhaustive. Further, we see from (45) that chains of class two are always bijective with their image under projection into the z-plane, while curves of class three are bijective only when B is an integer.

DEFINITION. Let γ be a chain and $\gamma(t)$ an arbitrary parametrization. Let $\eta: \gamma \to \mathbb{P}^1$ be any choice of the projective parametrization. If $\eta(t)$ covers $\mathbb{P}^1$ once for $a \leq t < b$ then we call $\{\gamma(t): a \leq t < b\}$ a *projective cycle*. If $\eta(t)$ covers $\mathbb{P}^1 - \{\text{one point}\}$ once for $-\infty < t < \infty$ then we call $\{\gamma(t): -\infty < t < \infty\}$ a *projective cycle*.

LEMMA 1. *Let γ be a chain of class one. Then γ as a point set is covered infinitely often by any of its projective cycles.*

PROOF. We have seen that the chains with $z(t) \equiv z_0$ are given by

$$w = w_0 e^{-2it} \qquad \text{(cf. (47))},$$

and thus have period π. But we have also seen that the projective parameter can be taken to be t itself. So one projective cycle corresponds to $-\infty < t < \infty$ and covers the chain an infinite number of times.

LEMMA 2. *Let γ be a chain of class two or three. Then the closed curve in the z-plane is the bijective image of any projective cycle.*

PROOF. As $\eta: \gamma \to \mathbb{P}$ covers $\mathbb{P}$ once, t varies by $\frac{2\pi}{|\mu|^2}$ and so $z(t)$ returns to its original value.

REMARK. We are not saying that the chain itself is covered by one projective cycle but only that the projection of the chain onto the z-plane is so covered.

We now consider $\mathscr{R}_B$ as an S^1-bundle over $\mathbb{C} - \{0\}$. The fiber is a chain of class one.

LEMMA 3. *Let P and Q be points in the same fiber. Then P and Q lie strictly within some projective cycle. If γ is any other chain containing P and Q, then no projective cycle of γ contains both P and Q.*

PROOF. We know that the fiber is a chain (of class one) and that the projective cycle is given by $-\infty < t < \infty$. So clearly P and Q are interior points of this projective cycle. Any chain of class two or three that starts at P can return to Q only after a projective cycle (since the image in the z-plane must close) and so P and Q cannot be within any projective cycle. And a chain of class four cannot return at all.

Now let $U \subset \mathscr{R}_B$ be an open set containing some fiber.

LEMMA 4. *Any CR diffeomorphism $\Phi: U \to V$, $V \subset \mathscr{R}_{B'}$, preserves fibers.*

PROOF. Let γ be a fiber in U. Then $\Phi(\gamma)$ is a closed chain in V. Since γ is covered in less than a projective cycle, the same is true for $\Phi(\gamma)$. But only chains of class one are covered in less than one projective cycle. So $\Phi(\gamma)$ is also a fiber.

We now are ready to prove Theorem 1. Let $U \subset \mathscr{R}_B$ be an open set containing a point P, the fiber through P, and one complete chain of class three through P.

LEMMA 5. *If there exists a CR diffeomorphism $\Phi: U \to V$, $V \subset \mathscr{R}_{B'}$, then $B \equiv B' \pmod{Z}$.*

REMARK. This is, of course, a stronger result than Theorem 1.

PROOF. Let γ be the chain in U of class three. The distinguished CR orientation (see Lemma 1.19) induces orientations on γ and $\Phi(\gamma)$, and Φ respects these orientations. In terms of the orientation of γ let $P, P_1, P_2, \ldots$

be the sequence of points along γ that are also on the fiber through P. Then $\Phi(P)$, $\Phi(P_1)$, $\Phi(P_2)$, ... is a sequence of points on the fiber through $\Phi(P)$. Because of (44), these sequences may be written as

$$w_n = w_{n-1}e^{2\pi i B} \quad \text{and} \quad w'_n = w'_{n-1}e^{2\pi i B'}.$$

In other words, Φ is an orientation preserving map of S^1 to itself that takes $e^{2\pi i B N}$ to $e^{2\pi i B' N}$ for all N. We wish to conclude that the existence of such a Φ implies $B \equiv B' \pmod{Z}$.

We can add or subtract any integers from B and B' without changing the sequence of points. Thus we may assume that $0 < B < 1$, $0 < B' < 1$. Let $\sigma(N)$ count the number of times the sequence

$$1, e^{2\pi i B}, e^{2\pi i (2B)}, e^{2\pi i (3B)}, \ldots, e^{2\pi i (NB)}$$

passes, or is equal to, the point 1. The quantity $\sigma(N)$ is well-defined since the orientation of S' is fixed. For example, for $B = \frac{1}{3}$ and the counterclockwise orientation we have

$$\sigma(0) = 1, \quad \sigma(1) = 1, \quad \sigma(2) = 1, \quad \sigma(3) = 2, \ldots$$

and in general

$$\sigma(3p + r) = p + 1 \qquad (r = 0, 1, 2),$$

while for $B = \frac{2}{3}$ we have

$$\sigma(0) = 1, \quad \sigma(1) = 1, \quad \sigma(2) = 2, \quad \sigma(3) = 3, \ldots$$

and in general

$$\sigma(3p) = 2p + 1,$$
$$\sigma(3p + 1) = 2p + 1,$$
$$\sigma(3p + 2) = 2p + 2.$$

Note that for these two examples

$$\lim_{N \to \infty} \frac{\sigma(N)}{N} = B.$$

LEMMA 6. *For any* B, $0 < B < 1$, *one has*

$$\lim_{N \to \infty} \frac{\sigma(N)}{N} = B.$$

REMARK. The result holds for any B, provided the definition of $\sigma(N)$ is modified in an obvious way.

PROOF. The quantity $\sigma(N)$ is also the number of zeros of $f(z) = e^{2\pi i B z} - 1$ in an appropriate domain. The argument principle, together with a simple residue calculation, can be used to establish the result.

With this, the proof of Lemma 5, and Theorem 1, is completed.

We conclude this section with the observation that the classes of chains introduced above can in fact be defined in a CR invariant way. That is, they can be defined using only the CR structure of the hypersurface rather than using the specific representation as a graph over the deleted z-plane.

PROPOSITION 3. *The classes defined above coincide, for B not an integer, with the classes:*

1. *Chains that close in less than one projective cycle.*
2. *Chains that close in one projective cycle.*
3. *Chains that do not close in one projective cycle.*
4. *Chains that do not close and that do not contain more than one projective cycle.*

When B is an integer, the third class is empty.

§3. **Chain-preserving maps.** For our next topic we start with the fact that a CR diffeomorphism preserves chains. We want to show that the converse is essentially true: A diffeomorphism between nondegenerate CR manifolds that maps the chains of one to the chains of the other is either a CR diffeomorphism or a conjugate CR diffeomorphism. This result is due to Cheng [**Ch**]. We first discuss conjugate structures and then give his proof in the three-dimensional case. Let (M, V) define a CR structure. Recall that this means that M is a manifold of dimension $2n + 1$ and V is a complex subbundle of the complexified tangent space with $\dim_{\mathbb{C}} V = n$ and $V \cap \overline{V} = \{0\}$. Note that $\overline{V}$ also satisfies these conditions and so $(M, \overline{V})$ also defines a CR structure. This one is called the conjugate CR structure (with respect to the original structure). The CR structure (M^3, V) can also be described, at least locally, by (ω, ω_1) where $\omega \wedge \omega_1 \wedge \overline{\omega}_1 \neq 0$ and $\omega^{\perp} \cap \omega_1^{\perp} = V$. So the conjugate structure is given by $(\omega, \overline{\omega}_1)$. Now we consider a strictly pseudoconvex CR structure.

EXERCISE 4. Start with a CR structure given by ω and ω_1, normalized so that $d\omega = i\omega_1 \overline{\omega}_1 \pmod{\omega}$. Let $\theta = -\omega$, $\theta_1 = \overline{\omega}_1$. Define, as in (6.50),

$$W = |\ell|^2 \theta,$$
$$W_1 = \ell(\theta_1 + u\theta),$$
$$W_2 = \frac{d\ell}{\ell} + \cdots,$$
$$W_3 = \frac{1}{\ell}(du + \cdots),$$
$$W_4 = \frac{1}{|\ell|^2}(dr + \cdots),$$

where $W, \ldots, W_4$ are the unique forms satisfying the structure equations (6.51). Show that in terms of the original CR structure we have

$$W = -\Omega,$$
$$W_1 = \overline{\Omega}_1,$$
$$W_2 = \overline{\Omega}_2,$$
$$W_3 = -\overline{\Omega}_3,$$
$$W_4 = -\Omega_4,$$

and $\lambda = \overline{\ell}$, $\mu = -\overline{u}$, $\rho = -r$.

Note that $\{W_1, \overline{W}_1, W_3, \overline{W}_3\}$ and $\{\Omega_1, \overline{\Omega}_1, \Omega_3, \overline{\Omega}_3\}$ generate the same differential ideal. The integral submanifolds coincide and, in particular, the projections into M coincide. Thus we have the following result.

PROPOSITION 4. *The CR structure (M, V) and its conjugate CR structure $(M, \overline{V})$ have the same chains.*

Recall that a map $\phi: (M, V) \to (M_1, V_1)$ is CR if $\varphi_* V \subset V_1$. We say that ϕ is conjugate CR if it provides a CR map of (M, V) to $(M_1, \overline{V}_1)$. That is, ϕ is conjugate CR if $\varphi_* V \subset \overline{V}_1$. If $\{\omega, \omega_1\}$ and $\{\theta, \theta_1\}$ give the CR structures of (M, V) and (M_1, V_1) then ϕ is CR if

$$\varphi^*(\theta) = \lambda\omega \quad \text{and} \quad \varphi^*(\theta_1) = U\omega_1 + v\omega,$$

and φ is conjugate CR if

$$\varphi^*(\theta) = \lambda\omega \quad \text{and} \quad \varphi^*(\theta_1) = V\overline{\omega}_1 + v\omega.$$

THEOREM 3. *Let $\phi: M \to \widetilde{M}$ be a diffeomorphism between strictly pseudo-convex three-dimensional CR structures. If ϕ maps the chains of M to the chains of $\widetilde{M}$ then ϕ is either a CR diffeomorphism or a conjugate CR diffeomorphism.*

PROOF. Since ϕ is a diffeomorphism, ϕ_* at each point p is a linear isomorphism of $T_p M$ to $T_{\phi(p)} \widetilde{M}$. Since ϕ preserves chains, ϕ_* maps each direction transverse to H_p to a direction transverse to $H_{\phi(p)}$. It follows that ϕ_* maps H_p to $H_{\phi(p)}$. Thus if $\{\omega, \omega_1\}$ gives the CR structure of M and $\{\widetilde{\omega}, \widetilde{\omega}_1\}$ gives the CR structure of $\widetilde{M}$, then

$$(49) \qquad\qquad \phi^*(\widetilde{\omega}) = \lambda\omega,$$

for some real nonzero function λ. Writing $\phi^*(\widetilde{\omega}_1)$ in terms of $\{\omega, \omega_1, \overline{\omega}_1\}$, we have

$$(50) \qquad\qquad \phi^*(\widetilde{\omega}_1) = U\omega_1 + V\overline{\omega}_1 + v\omega$$

for complex functions U, V, and v. We seek to show that either $V \equiv 0$ (and so ϕ is a CR map) or $U \equiv 0$ (and so ϕ is a conjugate CR map). For notational ease we let $\widetilde{\omega}$ also denote $\varphi^*\widetilde{\omega}$ and $\widetilde{\omega}_1$ also denote $\varphi^*\widetilde{\omega}_1$. Then we may think of $\{\omega, \omega_1\}$ and $\{\widetilde{\omega}, \widetilde{\omega}_1\}$ as defining CR structures on the same piece of $\mathbb{R}^3$, coordinatized by (x_1, x_2, x_3). Let $\gamma(t) = (x_1(t), x_2(t), x_3(t))$ be a chain with respect to the $\{\omega, \omega_1\}$ structure. Since ϕ preserves chains, $\gamma(t)$ is also a chain with respect to the $\{\widetilde{\omega}, \widetilde{\omega}_1\}$ structure. Thus $\gamma(t)$ satisfies the equations

$$\omega_1\left(\frac{d\gamma}{dt}\right) = -\mu\omega\left(\frac{d\gamma}{dt}\right),$$

$$\frac{d\mu}{dt} = (i\mu|\mu|^2 + B(\gamma(t), \mu(t)))\omega\left(\frac{d\gamma}{dt}\right),$$

and the same for ω, ω_1, μ, B replaced by $\widetilde{\omega}$, $\widetilde{\omega}_1$, $\widetilde{\mu}$, $\widetilde{B}$. Recall that the function $\mu(t)$ is defined by the first of these equations and that $|B| < C(1+|\mu|^2)$. In the realizable case (i.e., $d\omega_1 = 0$), this inequality comes from our basic equation (7). In the possibly nonrealizable case (where we might not be able to take $d\omega_1$ to be zero) the inequality still follows since the equation for the chains can differ from (7) only by terms independent of μ. See the remarks in Chapter 6 following the summary of Cartan's construction.

With our notational change we may rewrite (49) and (50) as

(51) $$\widetilde{\omega} = \lambda\omega \quad \text{and} \quad \widetilde{\omega}_1 = U\omega_1 + V\overline{\omega}_1 + v.$$

For the proof of the theorem it suffices to pick some point p and show that either $U(p) = 0$ or $V(p) = 0$. (Note that both equations cannot hold since ϕ is a diffeomorphism.)

We assume that chains in one CR structure are also chains in the other. Since the characteristic tangent spaces H coincide, a chain with $|\mu|$ large also has $|\widetilde{\mu}|$ large. From now on we restrict attention to chains that satisfy both

(52) $$|\mu| > 1 \quad \text{and} \quad |\widetilde{\mu}| > 1.$$

Along a chain we have

$$\omega_1 = -\mu\omega \quad \text{and} \quad \widetilde{\omega}_1 = -\widetilde{\mu}\omega.$$

From this and (51) we derive

(53) $$-\mu(t)U(\gamma(t)) - \overline{\mu}(t)V(\gamma(t)) + v(\gamma(t))$$
$$= -\widetilde{\mu}(t)\lambda(\gamma(t)).$$

Since λ is nonzero, there exists a constant C_1 independent of the particular chain such that

$$|\widetilde{\mu}| < C_1|\mu|,$$

provided $|\mu| \geq 1$. The same argument applied to ϕ^{-1} gives $|\mu| < C_2|\widetilde{\mu}|$ provided $|\widetilde{\mu}| \geq 1$. Thus there exist constants such that

$$C_2|\mu| < |\widetilde{\mu}| < C_1|\mu|,$$

provided (52) holds.

Next, we assume that each chain γ is parametrized so that $\left|\frac{d\gamma}{dt}\right| = 1$ at the point p. Here the norm is with respect to any fixed Riemannian metric. It is clear that there then exist some constants a and c such that at p

$$\frac{c}{|\mu|} < \left|\omega\left(\frac{d\gamma}{dt}\right)\right| < a,$$

provided $|\mu| \geq 1$.

We differentiate (53) and use

$$\frac{d\mu}{dt} = (i\mu|\mu|^2 + B(\gamma(t), \mu(t)))\omega(t),$$
$$\frac{d\widetilde{\mu}}{dt} = (i\widetilde{\mu}|\widetilde{\mu}|^2 + \widetilde{B}(\gamma(t), \widetilde{\mu}(t)))\omega(t)$$

to obtain, at p,

$$-(i\mu|\mu|^2 + B(\gamma,\mu))\omega\left(\frac{d\gamma}{dt}\right)U - (-i\overline{\mu}|\mu|^2 + \overline{B})\omega\left(\frac{d\gamma}{dt}\right)V$$

$$= -(i\widetilde{\mu}|\widetilde{\mu}|^2)\omega\left(\frac{d\gamma}{dt}\right)\lambda + O(|\mu|).$$

Here $O(|\mu|)$ denotes a term bounded by $C|\mu|$ where C is independent of the chain γ as long as γ has $|\mu| > 1$, all functions are evaluated at p, and we have used that $O(|\widetilde{\mu}|)$ can be replaced by $O(|\mu|)$. Further, since $B = O(|\mu|^2)$ and $\omega\left(\frac{d\gamma}{dt}\right) = O\left(\frac{1}{|\mu|}\right)$, we have

$$-i\mu|\mu|^2 U + i\overline{\mu}|\mu|^2 V = -i\widetilde{\mu}|\widetilde{\mu}|^2\lambda + O(|\mu|^2).$$

We replace $\widetilde{\mu}$ by its expression from (53) to obtain

$$-i\mu|\mu|^2 U + i\overline{\mu}|\mu|^2 V$$

$$= -\frac{i}{|\lambda|^2}(\mu U + \overline{\mu}V)(|\mu|^2|U|^2 + \mu^2 U\overline{V} + \overline{\mu}^2\overline{U}V + |\mu|^2|V|^2)$$

$$+ O(|\mu|^2).$$

Here U, V, and λ are evaluated at p and this is an identity in μ, i.e., it holds for each chain γ near enough to H. But then the coefficient of, for example, μ^3, on the right-hand side must be zero. Hence

$$U^2\overline{V} = 0,$$

and so either $U(p)$ or $V(p)$ is zero. This proves Theorem 3.

§4. Other results. Sometimes it is useful to have the chain equations written directly in local coordinates rather than in terms of b and the forms ω and ω_1. We now do this for a hypersurface in Moser normal form. As a side result of the necessary computations we will be able to relate the Cartan curvature r to the first nonzero coefficient $F_{42}(0)$ in the Moser normal form. We start with

$$(54) \qquad v = |z|^2 + \varphi(u)z^2\overline{z}^4 + \overline{\varphi(u)}\overline{z}^2 z^4 + O(|z|^7),$$

which we write as

$$(55) \qquad v = |z|^2 + F(z, \overline{z}, u).$$

We have from $(17')$

$$b = F_{zz\overline{z}} + O(|z|^5),$$

$$c = F_{zz\overline{z}\,\overline{z}} + O(|z|^4),$$

$$\ell = F_{zzz\overline{z}\,\overline{z}} + O(|z|^3).$$

For F as in (54) we thus have

$$b = 8\varphi(u)\overline{z}^3 + 24\overline{\varphi(u)}z^2\overline{z} + O(|z|^4),$$
$$c = 24(\varphi(u)\overline{z}^2 + \overline{\varphi(u)}z^2) + O(|z|^3),$$
$$\ell = 48\overline{\varphi(u)}z + O(|z|^2).$$

Here we digress for an important observation.

LEMMA 7. *Let M be in Moser normal form,*

$$v = |z|^2 + F_{24}(u)z^2\overline{z}^4 + F_{42}(u)z^4\overline{z}^2 + O(|z|^7),$$

and let

$$r = \frac{1}{6}(\ell_{\overline{1}} - 2\overline{b}\ell)$$

be the Cartan curvature computed with respect to the associated ω and ω_1. Then, at the origin,

$$r = 8F_{24}(0).$$

REMARK. The associated ω is given by (16), where r is the defining function not the curvature, and ω_1 is equal to dz.

PROOF. From a computation immediately preceding (17) we know that

$$\ell_1 = \ell_z + (i\overline{z} + O(|z|^5))\ell_u.$$

Since $b = 0$ at $z = 0$, we have

$$\overline{\ell}_{\overline{1}} = 48\varphi(u).$$

Thus, at the origin,

$$r = \frac{1}{6}(48\varphi(0)) = 8\varphi(0) = 8F_{24}(0).$$

We now return to our chain equations, which we may write as

$$dz = -\mu\omega,$$
$$d\mu = \{i\mu|\mu|^2 - (8\overline{\varphi}z^3 + 24\varphi\overline{z}^2z + O(|z|^4))|\mu|^2$$
$$+ 12i(\varphi(u)\overline{z}^2 + \overline{\varphi(u)}z^2 + O(|z|^3))\mu$$
$$- (8\varphi\overline{z} + O(|z|^2))\}\omega.$$

To write this in local coordinates we use (12):

$$\omega = \left(\frac{1}{2} + O(|z|^{12})\right)du + \left(\frac{-i}{2}\overline{z} + O(|z|^5)\right)dz$$
$$+ \left(\frac{i}{2}z + O(|z|^5)\right)d\overline{z}.$$

Let us introduce again the time parameter along a given chain for which

$$\omega\left(\frac{d}{dt}\right) = 1.$$

Thus

$$\frac{dz}{dt} = -\mu,$$

$$\frac{d\mu}{dt} = i\mu|\mu|^2 - \overline{b}|\mu|^2 + \frac{1}{2}ic\mu - \frac{1}{6}\overline{\ell},$$

and

$$\left(\frac{1}{2} + O(|z|^{12})\right)\frac{du}{dt} + \left(\frac{-i}{2}\overline{z} + O(|z|^5)\right)\frac{dz}{dt}$$

$$+ \left(\frac{i}{2}z + O(|z|^5)\right)\frac{d\overline{z}}{dt} = 1.$$

We may rewrite these equations as

(56)
$$\begin{cases} \dfrac{du}{dt} = 2 - i\overline{z}\mu + iz\overline{\mu} + A, \\[2mm] \dfrac{dz}{dt} = -\mu, \\[2mm] \dfrac{d\mu}{dt} = i\mu|\mu|^2 - (8\overline{\varphi}z^3 + 24\varphi\overline{z}^2 z)|\mu|^2 \\[2mm] \qquad\qquad + 24i(\mathrm{Re}(\varphi\overline{z}^2))\mu - 8\varphi\overline{z} + B, \end{cases}$$

with

$$|A| \le C|\mu|O(|z|^5)$$

and

$$|B| \le C(|z|^2 + |z|^3|\mu| + |z|^4|\mu|^2).$$

In some situations one wishes to study chains along which $|\mu| \to \infty$. Here it is useful to introduce another time parametrization. So let s and t be related by

(57)
$$\frac{dt}{ds} = \frac{1}{|\mu(t)|^2}$$

along a given chain (provided this chain always has μ different from zero). In fact, we used this parametrization at the start of this chapter when we derived the explicit representation of the chains on Q.

The equations now become ($\dot{x}$ stands for $\frac{dx}{ds}$).

(58)
$$\begin{cases} \dot{u} = \dfrac{2}{|\mu|^2} + \dfrac{iz}{\mu} - \dfrac{i\overline{z}}{\overline{\mu}} + a, \\[2mm] \dot{z} = -\dfrac{1}{\overline{\mu}}, \\[2mm] \dot{\mu} = i\mu - 8(\overline{\varphi}z^3 + 3\varphi|z|^2\overline{z}) + 24i\mathrm{Re}(\varphi\overline{z}^2)\left(\dfrac{1}{\overline{\mu}}\right) \\[2mm] \qquad - 8\varphi z\left(\dfrac{1}{|\mu|^2}\right) + b, \end{cases}$$

with

$$|a| = |A|/|\mu|^2,$$
$$|b| = |B|/|\mu|^2.$$

It is natural to think of the chains in CR geometry as the analog of the geodesics in Riemannian geometry. Certainly, both families of curves are given as the solutions to a second-order equation. And it has been shown ([**Ja2**], [**Koc1**], [**Kr**]) that any two nearby points on a strictly pseudoconvex CR hypersurface may be connected by a chain. (This result is false if strictly pseudoconvex is replaced by nondegenerate although there is a generalized family of chains for which the result still holds [**Koc2**].) But chains differ from geodesics in at least two important ways. First, it is not true that any two arbitrary points can always be connected by a chain.

EXAMPLE ([**BS**]). Let (z, u) be the usual coordinates on Q. The map $(z, u) \to (rz, r^2 u)$ is a CR diffeomorphism for any fixed r different from zero. When r is also different from one, this map generates an infinite cyclic group G of diffeomorphisms. Let $\mathscr{D} = Q - \{0\}/G$ denote the quotient under the equivalence relation $(z, u) \sim (r^n z, r^{2n} u)$, $n \in \mathbb{Z}$. $\mathscr{D}$ clearly is a smooth manifold and has a CR structure that is locally the same as that of Q. In particular, any chain of Q maps to a chain of $\mathscr{D}$ and all chains of $\mathscr{D}$ are so obtained. Let $\pi: Q - \{0\} \to \mathscr{D}$ be the projection. So $\pi(z, u) = [z, u] = \{(\zeta, \eta) = (r^n z, r^{2n} u)$ for some $n \in \mathbb{Z}\}$. We claim there is no chain connecting the points $[0, 1]$ and $[0, -1]$. For if such a chain, say γ, existed, there would be a chain Γ in Q that connects $(0, 1)$ and $(0, -r^{2n})$ for some n and for which $\Gamma - \{0\}$ projects to γ. But the only chain in Q between these two points is $\Gamma = \{(0, u): u \in \mathbb{R}\}$ and $\pi(\Gamma - \{0\})$ consists of two disjoint closed curves in $\mathscr{D}$ and so cannot be γ.

EXERCISE 5. Show that $\mathscr{D}$ is compact and connected.

The second difference is more surprising since it is local. Fefferman [**Fef**] has shown that the hypersurface

$$v = |z|^2 + u|z|^8$$

has a family of chains that "spiral" into the origin. His proof is too long (and relies on a completely different approach to chains) to include here. So instead we make some elementary remarks indicating that in some sense spiraling is the only type of singularity to be expected. Let γ be a chain with some parametrization t, $-\infty < t < \infty$, and let p be a limit point in the sense that

(a) $\lim_{t \to \infty} \gamma(t) = p$,

(b) there does not exist a chain $\tilde{\gamma}$ containing p as an interior point and also containing the chain γ.

This definition is clearly independent of the choice of parameter (as long as orientation is preserved). Note that we are not considering chains with a more general limit point where $\lim \gamma(t) = p$ is required only for some sequence $t_j \to \infty$.

We first show that $\gamma(t)$ must become more "horizontal" as it approaches p.

LEMMA 8. $\lim_{t \to \infty} |\mu(t)| = \infty$.

PROOF. For each point q and each initial value μ_0, there is some $\varepsilon(q, \mu_0)$ for which the unique chain through q and having "slope" μ_0 at q exits $B(\varepsilon, q)$. By "exits" we mean in each direction along the chain. By an elementary result in ordinary differential equations, the function $\varepsilon(q, \mu_0)$ can be chosen to be continuous. Thus if q is restricted to a compact neighborhood of p and μ_0 is restricted so that $|\mu_0| \le C$ for some constant C, then the chain determined by q and μ_0 exits some ball $B(\varepsilon, q)$ where ε is independent of q and μ_0.

To prove the lemma, assume on the contrary that there is a sequence $t_j \to \infty$ and a constant C such that $|\mu(t_j)| \le C$. Choose a particular j for which $\text{dist}(p, \gamma(t)) < \varepsilon/2$ for all t, $t > t_j$, where ε is as above. We now have a contradiction. For the chain through $q = \gamma(t_j)$ with slope $\mu_0 = \mu(t_j)$ must exit $B(\varepsilon, q)$. And in order to do this it must exit $B(\varepsilon/2, p)$. But $\{\gamma(s): t > t_j\}$ can never leave $B(\varepsilon/2, p)$.

So we now consider a chain, together with a more-or-less arbitrary parametrization, for which $\lim_{t \to \infty} \gamma(t) = p$ and $\lim_{t \to \infty} |\mu(t)| = \infty$. There is of course the CR invariant projective parametrization which, however, is not so useful here. More interesting is the "natural" but not invariant parametrization by arc length, which uses the Euclidean metric on $\mathbb{C}^2$. We want to show that γ must have infinite arc length near a limit point. (This result was obtained independently by both J. Bland and G. Toth.) Our tool for doing this is the parametrization given by (57). We start with the equations (58). We consider z, u, and μ as functions of s along the given chain. We also take p to be the origin. Thus from the chain equations we derive

(59)
$$\dot{z} = -\frac{1}{\mu},$$

$$\dot{\mu} = i\mu + g(s),$$

where $g(s)$ approaches zero as the point on the chain approaches the origin. Now, either our parameter s approaches some finite value as $\gamma(s)$ approaches the origin or our parameter approaches infinity. If s only approaches some finite value, then the second equation implies that $|\mu|$ remains finite. So we must have that $s \to \infty$ as the chain approaches its singularity.

Indeed, the second equation of (59) can be solved explicitly,

$$\mu(s) = C_1 e^{is} + e^{is} \int_{s_0}^{s} e^{-i\sigma} g(\sigma) \, d\sigma.$$

Hence

$$|\mu(s)| < C_1 + C_2 s,$$

and, in particular,

$$\int_{s_0}^{\infty} \frac{1}{|\mu(s)|} \, ds = \infty.$$

LEMMA 9. *Let γ be a chain on $M^3 \subset \mathbb{C}^2$ where $\mathbb{C}^2$ has the usual Euclidean structure of $\mathbb{R}^4$. If γ has a limit point (as defined above) then the arc length of γ is infinite.*

REMARK. The arc length of a chain is clearly not a CR invariant but the property of having finite (or infinite) arc length is invariant under any diffeomorphism. Thus we may assume that M is in partial normal form.

PROOF. Consider the curve in the z-plane that is defined by (59). This curve is a projection of the chain and so it is sufficient to show that this curve has infinite arc length. This is so since

$$\int_{s_0}^{\infty} \left| \frac{dz}{ds} \right| \, ds = \int_{s_0}^{\infty} \frac{1}{|\mu(s)|} \, ds = \infty.$$

It is also easy to see that the curve $z(t)$ is locally convex in the sense that for each t_0 there is some ε_0 for which the curve segment $\{z(t): t_0 - \varepsilon < t < t_0 + \varepsilon\}$ lies on one side of the tangent line to the curve $z(t)$ at t_0 and touches this tangent line only at t_0.

EXERCISE 6.

(a) Show that a curve in $\mathbb{C}^1$ is convex near t_0 if at t_0

$$\text{Im} \left(\frac{d^2 z}{dt^2} \frac{d\bar{z}}{dt} \right) \neq 0.$$

(b) Use this to show that the projection of any chain is locally convex near a limit point.

CHAPTER 9

Chains and Circles in Complex Projective Geometry

§1. Circles and anti-involutions. As is easily imagined, Cartan's motivation for introducing chains was not to notice that $\{\Omega_1, \Omega_3\}$ happens to define a closed differential ideal whose integral elements happen to project down only to curves. Rather, there is in complex projective geometry a classical object called a chain. This was introduced by von Staudt [St]; see also [Ca1] and [NV].

Given four points of $\mathbb{CP}^1$, $Z_j = [z_j, w_j]$, $j = 1, \ldots, 4$, the harmonic ratio is defined as

$$(Z_1, Z_2, Z_3, Z_4) = \frac{(x_1 - x_3)(x_2 - x_4)}{(x_1 - x_4)(x_2 - x_3)},$$

where $x_j = w_j/z_j$ takes its values in the extended complex plane.

DEFINITION. A subset Γ of $\mathbb{CP}^1$ is called a *chain* if the harmonic ratio of four arbitrary points in Γ is always real.

We have the usual inclusion $\mathbb{C} \subset \mathbb{CP}^1$. It is not hard to see that a chain must be either a real line or a circle in $\mathbb{C}$. Thus a map that takes chains to chains must be projective or the conjugate of a projective map. In particular, a chain-preserving map is holomorphic or antiholomorphic (cf. Theorem 8.3).

A chain also has the property that it is the set of real points in some holomorphic coordinate system. Thus it is the fixed point set of the conjugation operator in such a coordinate system. Another way of saying this is that there is an antiholomorphic map $A: \mathbb{C} \to \mathbb{C}$ with $A^2 = \text{Id}$ whose fixed point set is Γ.

DEFINITION. An *anti-involution* on $\mathbb{CP}^n$ is an antiholomorphic projective map of $\mathbb{CP}^n$ to itself whose square is the identity. An *anti-involution* on $\mathbb{C}^n$ is the map induced by an anti-involution on $\mathbb{CP}^n$.

REMARK. In the usual homogeneous coordinates for $\mathbb{CP}^n$ an anti-involution is given by

$$Z \to A\overline{Z},$$

where A is a matrix satisfying

$$A\overline{A} = \text{Id}.$$

189

The hyperquadric $Q \subset \mathbb{C}^2$ has the property that its intersection with a complex line produces within that complex line either a real line or a circle. So Cartan naturally called such intersections chains and then set for himself the problems of finding the equations of the chains with respect to arbitrary coordinates on Q and of generalizing the idea of chains to other hypersurfaces. To see how he might have proceeded, recall our description in Chapter 2 of $\mathrm{SU}(2, 1)$ in terms of matrices $A = (A_0, A_1, A_2)$. Here A_0 and A_2 are the homogeneous coordinates of distinct points in Q and A_1 is the homogeneous coordinate of the polar point for the complex line determined by $[A_0]$ and $[A_2]$. We hold A_0 fixed and allow $[A_2(t)]$ to trace out a piece of a chain through A_0 (but $[A_2(t)]$ is never equal to $[A_0]$). Thus $[A_1]$ remains fixed. From

(1) $$dA_1 = \Omega_{1k} A_k,$$

we see that Ω_{10} and Ω_{12} are zero. But from (5.21) and the equations preceding it,

(2) $$\Omega_{10} = -i\overline{\Omega_{21}} = i\overline{\Omega_3},$$
$$\Omega_{12} = i(\overline{\Omega_{01}}) = i\overline{\Omega_1},$$

so $\{\Omega_1, \Omega_3\}^{\perp}$ defines the totality of chains through a given point. We would now only have to verify that $\{\Omega_1, \Omega_3\}$ is a closed differential ideal on any M^3 and that the projection is indeed a curve in order to have our generalization of chains to arbitrary strictly pseudoconvex hypersurfaces. This, of course, is what we did in the previous chapter.

Cartan also used a concept from classical projective geometry to define a second set of curves on an arbitrary strongly pseudoconvex M^3. These are his "circles" which we shall refer to as "pseudocircles."

The subject begins with a particular curve on S^3, namely the circle

$$\mathscr{C}_0 : x^2 + u^2 = 1.$$

Here the coordinates for $\mathbb{C}^2$ are $z = x + iy$ and $w = u + iv$. Thus we are looking at the intersection of S with the totally real plane $\operatorname{Im} z = \operatorname{Im} w = 0$. Note that this circle is the intersection of S with the fixed point set of an anti-involution of $\mathbb{C}^2$. In classical notation, the set of fixed points of an anti-involution on $\mathbb{CP}^{k-1}$ is called a $(k-1)$-chain. So our chains are 1-chains and our circles are the intersection of S with 2-chains (but not with arbitrary 2-chains, see below).

We now consider the family of all images of $\mathscr{C}_0$ under automorphisms of S. Since the group of automoprhisms is of dimension 8 while the subgroup of those that map $\mathscr{C}_0$ to itself is of dimension 3 (see the second corollary to Theorem 2.5), this family of images is of dimension 5.

EXERCISE 1. There is a three-parameter family of images that go through a given point. (Hint: This comes from a dimension count.)

DEFINITION. A *pseudocircle* in S^3 is any image of the curve

(3) $$\mathscr{C}_0: x^2 + u^2 = 1, \quad y = 0 = v,$$

under a CR automorphism of S^3.

EXERCISE 2. The tangent to a pseudocircle is always in H, the characteristic two-plane distribution.

Thus, in some sense, the pseudocircles are complementary to the chains, which are never in the H direction. However, as we see from the above exercises, each direction in H must define not just one pseudocircle but a two-parameter family of pseudocircles. Later we shall see this more directly.

We shall also work with the pseudocircles on

$$Q = \left\{ (z_1, z_2): \operatorname{Im} z_2 = \frac{1}{2} |z_1|^2 \right\}.$$

These again are defined as the images of some standard curve. The curve we take is the restriction to Q of

$$z_1^2 = -2z_2.$$

In parametric form, the curve is given by

(4) $$z_1 = (1 - i)t, \quad z_2 = it^2.$$

The reason for this choice should become clear later. Naturally we want the pseudocircles on S to map to those on Q (with the usual caution about the point at infinity). This is an easy fact. See Exercise 2.4.

Now, $\mathscr{C}_0$ is certainly the intersection of S^3 with a real 2-plane. Just as certainly, not all pseudocircles on S^3 arise in this way since complex projective maps do not preserve real planes (see the discussion preceding Lemma 2.8). But every pseudocircle is the intersection with a plane when we go to homogeneous coordinates.

LEMMA 1. *Every pseudocircle on* S^3 *is the projection of the intersection of* $\tilde{S} \subset \mathbb{C}^3$ *with a totally real 3-plane.*

NOTATION. Throughout this chapter we use $[z]$ to denote the projection of $\mathbb{C}^3$ to $\mathbb{CP}^2$ or to $\mathbb{C}^2$, $[\Phi]$ to denote the map of $\mathbb{CP}^2$ or $\mathbb{C}^2$ induced, via this projection, by the linear map $\Phi: \mathbb{C}^3 \to \mathbb{C}^3$, and $\tilde{S}$ and $\tilde{Q}$ to denote the lifts of S and Q to $\mathbb{CP}^2$. That is, in homogeneous coordinates,

$$\tilde{S} = \{ [z_0, z_1, z_2]: |z_1|^2 + |z_2|^2 - |z_0|^2 = 0 \}$$

and

$$\tilde{Q} = \{ [z_0, z_1, z_2]: -i|z_0|^2 + i|z_2|^2 - |z_1|^2 = 0 \}.$$

PROOF. Let P denote the real 3-plane $\{ x_0, x_1, x_2 \mid x_j \text{ is real} \}$ in $\mathbb{C}^3$. So

$$\mathscr{C}_0 = [\tilde{S} \cap P].$$

For any other pseudocircle $\mathscr{C}$ there is an automorphism ϕ of S^3 with $\phi(\mathscr{C}_0) = \mathscr{C}$. Lift ϕ to an element Φ of $SU(2, 1)$. That is, $[\Phi(Z)] = \phi[Z]$. Since Φ is linear and preserves $\tilde{S}^3$ we have that

$$\Phi(\tilde{S}^3 \cap P) = \tilde{S}^3 \cap P'$$

for some plane P' and so

$$\mathscr{C} = \phi(\mathscr{C}_0) = [\tilde{S}^3 \cap P'].$$

P' is totally real since it is the holomorphic image of the totally real set P.

REMARKS. 1. If this implied that every pseudocircle is the intersection of S^3 with a 2-plane, then we would have a contradiction. But this is not implied since the projection of a real plane in $\mathbb{C}^3$ is not in general a plane in $\mathbb{C}^2$.

2. Not every totally real 3-plane gives rise to a pseudocircle. Soon we will characterize those that do.

3. The plane is not unique: Let P be totally real and let μ be an arbitrary complex number different from zero. The plane $P' = \{\mu Z : Z \in P\}$ is also totally real and has the same projection as P. In fact, these are the only planes that have the same projection as P.

EXERCISE 4. (a) Let P and P' be totally real 3-planes with $[P] = [P']$. Show that there exists some μ with $P' = \mu P$. (Hint: For each $X \in P$ there exists some μ such that $\mu X \in P'$. Show that μ does not depend on X.)

(b) Let β be a nonzero complex number. Show that the real 3-plane P spanned by the vectors

$$(1, 0, 1), \qquad (0, \beta, 0), \qquad (i, 0, -i)$$

is totally real. Then show that $[P \cap \tilde{S}]$ does not depend on β. Thus from $[P \cap \tilde{S}] = [P' \cap \tilde{S}]$ we cannot conclude that $[P] = [P']$. This observation plays a role in the proof of Theorem 1.

There is a correspondence between totally real n-planes and anti-involutions in $\mathbb{C}^n$. Before explaining this, we look at two examples. The simplest anti-involution is of course conjugation with respect to the usual basis of $\mathbb{C}^n$. In $\mathbb{C}^3$ this map is

$$A(z_0, z_1, z_2) = (\bar{z}_0, \bar{z}_1, \bar{z}_2).$$

Its fixed point set is the plane $\{\text{Im } z_0 = \text{Im } z_1 = \text{Im } z_2 = 0\}$. Note that as a map on $\mathbb{C}^2$, A maps S^3 to itself and its fixed point set in S^3 is $\mathscr{C}_0$.

Another anti-involution, important to us later, is

$$B(z_0, z_1, z_2) = (i\bar{z}_0, \bar{z}_1, -i\bar{z}_2).$$

This map preserves $\tilde{Q}$, and its fixed point set in $\tilde{Q}$ is

(5) $$\{(1 + i)u^2, \ 2uv, \ -(1 - i)v^2\}.$$

As a map on $\mathbb{C}^2$, B maps Q to Q and its fixed point set is the pseudocircle (4).

In some sense A is typical—each anti-involution is given by conjugation. We want to explain this and then to describe the fixed point set of any anti-involution. Recall the relation of $\mathbb{R}^6$ and $\mathbb{C}^3$ as in Chapter 1. The map of the underlying spaces is given by $X \to X + iJX$. A 3-plane in $\mathbb{R}^6$ is totally real if it has a basis $\{X_1, X_2, X_3\}$ such that $\{X_1, X_2, X_3, JX_1, JX_2, JX_3\}$ is a basis for $\mathbb{R}^6$. A 3-plane in $\mathbb{C}^3$ is totally real if it is the real span of some vectors $\{Z_1, Z_2, Z_3\}$ whose complex span is $\mathbb{C}^3$. There are two corresponding ways to obtain anti-involutions.

1. Given a totally real plane $\{X_1, X_2, X_3\}$ define a real linear map by

$$BX_k = X_k, \qquad BJX_k = -JX_k.$$

Clearly $B^2 = \mathrm{Id}$.

2. Given three independent vectors $\{Z_1, Z_2, Z_3\}$ in $\mathbb{C}^3$ define a map by

$$C\left(\sum z_j Z_j\right) = \sum \overline{z}_j Z_j.$$

Clearly C is an anti-involution. The conjugation A above is just a map of this form where $\{Z_1, Z_2, Z_3\}$ is the standard basis.

EXERCISE 5. (a) Show that B is antiholomorphic as a map on $\mathbb{C}^3$ and hence is an anti-involution. (Recall that to be antiholomorphic is equivalent to anticommuting with J.)

(b) Show that B and C are the same map when for Z_j we take $(X_j)_{\mathbb{C}}$. That is, show that for all $Y \in \mathbb{R}^6$, $(BY)_{\mathbb{C}} = CY_{\mathbb{C}}$.

Note that for B each point in $\{X_1, X_2, X_3\}$ is a fixed point and that for C each point in the real plane spanned by $\{Z_1, Z_2, Z_3\}$, that is, each real linear combination of these vectors, is a fixed point and that these are the only fixed points. Clearly the fixed points of any anti-involution form a real plane since any real linear combination of fixed points is still a fixed point. Note that if X is a fixed point of the anti-involution A, that is $A(X) = X$, then in matrix notation $A\overline{X} = X$.

LEMMA 2. *Let P be a totally real n-plane in $\mathbb{C}^n$. There is a unique anti-involution that has P as its fixed point set. Conversely, the fixed point set for any anti-involution on $\mathbb{C}^n$ is a totally real n-plane.*

PROOF. Starting with the totally real plane we define, using 1) or 2) as above, an anti-involution having that plane as its fixed point set. This anti-involution is unique because a holomorphic (or an antiholomorphic) mapping is completely determined by its values on a maximally real set.

Conversely, let us start with an anti-involution. We know that its fixed point set is a real plane and we need to show that it is totally real and of dimension n. Because $A\overline{A} = \mathrm{Id}$, it is easily seen that the Jordan canonical form for A can consist only of blocks of length one or two. In the first case $|\lambda| = 1$ and in the second $\lambda = \pm i$. Let X be a fixed point of A, that is, a solution to $A\overline{X} = X$. The components of X relative to the basis realizing the Jordan canonical form are

(6) $\quad (a_1 e^{i\theta_1}, \ldots, a_K e^{i\theta_K}, B_{K+1}(1 + A_{K+1}i), B_{K+1}(1 - A_{K+1})(1 \pm i),$
$\quad\quad \ldots, B_{K+P}(1 + A_{K+P}i), B_{K+P}(1 - A_{K+P})(1 \pm i)),$

where the corresponding eigenvalues are $\lambda_j = e^{2i\theta_j}$, $\lambda_{K+j} = \pm i$, the coefficients a_j, A_{K+j}, B_{K+j} are arbitrary real numbers, and $K + 2P = n$. Thus the set of fixed points is of real dimension n. That it is totally real comes from Exercise 1.5.

We know that two totally real 3-planes can have the same projection. What can we say about their associated anti-involutions? Recall that if P is a plane in $\mathbb{C}^3$ then $[P]$ is its projection into $\mathbb{C}^2$ (which need not be a plane) and if A is a linear or conjugate linear map of $\mathbb{C}^3$ then $[A]$ is the induced projective map on $\mathbb{C}^2$.

LEMMA 3. *Let P and P' be totally real 3-planes and let A and A' be their associated anti-involutions. Then $[P] = [P']$ if and only if $[A] = [A']$.*

PROOF. If $[A] = [A']$ then $A = \lambda A'$. (This is easiest seen by reducing to the case where A' is the identity matrix.) Since A and A' are anti-involutions we have that $|\lambda| = 1$. Hence if X is a fixed point of A', i.e., $A'X = X$, then $\overline{\mu}X$ is a fixed point of A, where $\lambda = \overline{\mu}/\mu$. Thus $[P] = [P']$.

Conversely, if $[P] = [P']$ then we know from Exercise 4 that $P' = \mu P$. Let $P = \{X_1, X_2, X_3\}$. Then

$$A'(z_j X_j) = A'\left(\frac{z_j}{\mu}\mu X_j\right) = \frac{\overline{z_j}}{\overline{\mu}}\mu X_j = \frac{\mu}{\overline{\mu}}A(z_j X_j).$$

That is,

$$A' = \left(\frac{\mu}{\overline{\mu}}\right)A,$$

and so $[A'] = [A]$.

We now look for those anti-involutions that map S to itself. If $A(Z)$ denotes an anti-involution and $A = (a_{ij})$ the corresponding matrix then

$$A^2(Z) = Z, \quad A\overline{A} = \mathrm{Id}, \quad A(z_1, \ldots, z_n) = a_{ij}\overline{z}_j, \quad a_{ij}\overline{a}_{jk} = \delta_{ik}.$$

The condition for a linear map A to take some hypersurface $F(Z, Z) = 0$ to itself is that

(7) $$A^*\overline{C}A = \lambda C,$$

where A^* is the adjoint (conjugate transpose) of the matrix A and C is the Hermitian matrix of the quadratic form F, i.e., $F(Z, W) = Z \cdot C\overline{W}$. We claim that if A is an anti-involution on an odd-dimensional space, then λ must be equal to one. For, from (7) and using $A\overline{A} = \mathrm{Id}$, we may derive

$$C = \lambda A^*\overline{C}A,$$

and so $\lambda^2 = 1$. But taking determinants in (7) yields $\lambda^n = 1$. When n is odd we have $\lambda = 1$. Thus in $\mathbb{C}^3$,

$$(8) \qquad\qquad A^* \overline{C} A = C$$

is the condition for an anti-involution to preserve the hypersurface $F(Z, Z) = 0$.

For $\tilde{S}$ we have $F = z_1 \overline{w}_1 + z_2 \overline{w}_2 - z_0 \overline{w}_0$ and

$$C = \begin{pmatrix} -1 & 0 & 0 \\ 0 & 1 & 0 \\ 0 & 0 & 1 \end{pmatrix}.$$

For $\tilde{Q}$ we have $F = -i z_0 \overline{w}_2 + i z_2 \overline{w}_0 + z_1 \overline{w}_1$ and

$$C = \begin{pmatrix} 0 & 0 & -i \\ 0 & 1 & 0 \\ i & 0 & 0 \end{pmatrix}.$$

Let P be a totally real plane whose associated anti-involution maps $\tilde{Q}$ to itself. Fix some μ with $\mu \neq 0$ and let

$$P' = \{\mu Z : Z \in P\}.$$

Then P' is also totally real and from (8) we see that its anti-involution also preserves $\tilde{Q}$.

We know that every pseudocircle is the projection of some totally real plane in $\mathbb{C}^3$ intersected with $\tilde{S}$. But not all totally real planes give pseudocircles. For instance, the projection might not be tangent to H. The next two theorems determine which totally real planes do give rise to pseudocircles.

THEOREM 1. *The projection of the intersection of $\tilde{S}$ with a totally real 3-plane is a pseudocircle if and only if the anti-involution determined by the plane maps $\tilde{S}$ to itself.*

THEOREM 2. *Let P be a totally real 3-plane and let $\{A_1, A_2, A_3\}$ be any basis for P. The anti-involution associated to P maps $\tilde{S}$ to itself if and only if each $F(A_j, A_k)$ is real.*

REMARKS. 1. It follows from the first theorem that if $[P \cap \tilde{S}] = [P' \cap \tilde{S}]$ then the anti-involutions associated to both P and P' preserve $\tilde{S}$ provided one of them does.

2. If the condition in Theorem 2 holds for one basis it holds for all bases. For the vectors in any other basis are real linear combinations of the vectors in the first basis and so $F(B_j, B_k)$ is also real.

3. For convenience these results are stated for $\tilde{S}$. They also hold for $\tilde{Q}$. This is true for Theorem 1 because there is a projective map of S to Q, and hence a linear map of $\tilde{S}$ to $\tilde{Q}$, while the proof of Theorem 2 holds for any Hermitian form F.

4. Let F be the bilinear form associated to $\tilde{S}$. The three-dimensional subspaces on which F is real give rise to the anti-involutions that preserve $\tilde{S}$.

EXAMPLES. For our original $\mathscr{C}_0$ we can take P_0 to be $\{(z_0, z_1, z_2): \text{each } z_j \text{ is real}\}$. Then the corresponding anti-involution is

$$A_0(z_0, z_1, z_2) = (\bar{z}_0, \bar{z}_1, \bar{z}_2).$$

This maps $\tilde{S}$ to itself. The bilinear form associated to $\tilde{S}$ is clearly real on P_0.

For the pseudocircle (4) on Q we can take A to be

$$A(z_0, z_1, z_2) = (i\bar{z}_0, \bar{z}_1, -i\bar{z}_2).$$

This preserves the quadratic form

$$F(Z, W) = -iz_0\bar{w}_2 + iz_2\bar{w}_0 + z_1\bar{w}_1,$$

and so maps $\tilde{Q}$ to itself. The fixed point plane of A is spanned by $\{(1 + i, 0, 0), (0, 0, 1 - i), (0, 1, 0)\}$ (see (5)) and it is easily checked that $F(A_j, A_k)$ is real for A_j and A_k any of these vectors.

PROOF OF THEOREM 1. Let P be the totally real plane, A the associated anti-involution and set $\mathscr{C} = [P \cap \tilde{S}]$. Similarly, we have P_0, A_0, and $\mathscr{C}_0$ where $\mathscr{C}_0$ is also given by (3). First let us assume that $\mathscr{C}$ is a pseudocircle. So there is an automorphism ϕ of S^3 with $\phi(\mathscr{C}) = \mathscr{C}_0$. This automorphism lifts to Φ, that is, $[\Phi] = \phi$. Since Φ is a linear map of $\tilde{S}$ to itself, we have

$$\Phi(P) \cap \tilde{S} = \Phi(P \cap \tilde{S}).$$

So

$$\mathscr{C}_0 = \phi(\mathscr{C}) = \phi(P \cap \tilde{S}) = [\Phi(P \cap \tilde{S})] = [\Phi(P) \cap \tilde{S}].$$

Thus $[(\Phi P) \cap \tilde{S}] = [P_0 \cap \tilde{S}]$. For arbitrary totally real planes this is not enough to conclude that $[\Phi P] = [P_0]$ (see Exercise 4). However, here $P_0 \cap \tilde{S}$ clearly contains three vectors independent over $\mathbb{C}$, e.g., $Z_1 = (1, 0, 1)$, $Z_2 = (1, 1, 0)$, $Z_3 = (\sqrt{2}, 1, 1)$. Thus $\Phi(P) \cap \tilde{S}$ contains $\mu_1 Z_1$, $\mu_2 Z_2$, $\mu_3 Z_3$ for certain complex constants μ_1, μ_2, μ_3. We seek to show that μ_j/μ_k is real, for then it follows that $\Phi(P) = \mu P_0$ and so $[\Phi(P)] = [P_0]$. An arbitrary point of $\Phi(P) \cap \tilde{S}$ is given by $W = \sum \lambda_j \mu_j Z_j$ where the λ_j's are real and satisfy some equation. We have that $[W] = [Z]$ for some point $Z \in P_0 \cap \tilde{S}$. There are real constants a_j such that $Z = \sum a_j Z_j$ and so there must be a complex number σ satisfying

$$\sigma \lambda_j \mu_j = a_j \quad \text{for} \quad j = 1, 2, 3.$$

It follows that μ_j/μ_k is real. The anti-involution $\Phi A \Phi^{-1}$ has ΦP as its fixed point set and so, because of Lemma 3, $[\Phi A \Phi^{-1}] = [A_0]$. Since $[\Phi]$ and $[A_0]$ map S^3 to itself, so does A.

Next let us assume that A maps $\tilde{S}$ to itself. We seek to find some $\phi \in$ SU(2, 1) such that $\phi(\mathscr{C}_0) = \mathscr{C}$. This is equivalent to finding some $\Phi \in$ SU(2, 1) that maps P_0 to P.

At this stage it is convenient to work with $\tilde{Q}$ rather than $\tilde{S}$ since our work with SU(2, 1) has predominantly used the action of this group on Q rather than on S. So we seek to show that an anti-involution that maps $\tilde{Q}$ to itself gives rise to a pseudocircle. To do this, we shall assume the validity of Theorem 2. (Note that the proof below of Theorem 2 does not rely on Theorem 1!) So for any basis $\{A_1, A_2, A_3\}$ of P we have that $F(A_j, A_k)$ is real. We want to construct an element Φ of SU(2, 1) that takes P_0 to P. As a basis for P_0 we may take $\{B_1, B_2, B_3\}$ with $B_1 = (1 + i, 0, 0)$, $B_2 = (1 + i, 2, -(1 + i))$, and $B_3 = (0, 0, -(1 - i))$. For P we choose any basis with A_1 and A_3 belonging to $\tilde{Q}$. We may find real numbers r and s so that $V = A_2 - rA_1 - sA_3$ satisfies

$$F(A_1, V) = 0 \quad \text{and} \quad F(A_3, V) = 0.$$

Recall that P and μP project to the same curve in Q and so we have the freedom to replace P by μP. We choose μ so that

$$\mu^3 \det(A_1, A_2, A_3) \quad \text{is real}$$

and then replace P by μP and A_j by μA_j. By the renaming and renumbering we end up with

$$P = \{A_0, A_1, A_2\},$$
$$0 = F(A_0, A_0) = F(A_2, A_2) = F(A_0, A_1) = F(A_2, A_1),$$
$$\det(A_0, A_1, A_2) \quad \text{is real}.$$

We now look for an element of SU(2, 1) that takes

$$\begin{aligned}
(1, 0, 0) &\quad \text{to} \quad \frac{1}{1 + i}A_0, \\
(0, 1, 0) &\quad \text{to} \quad \lambda_1 V,
\end{aligned}$$

and

$$(0, 0, 1) \quad \text{to} \quad \frac{\lambda_2}{1 - i}A_2.$$

Note that such an element would map P_0 to P provided λ_1 and λ_2 are real. So for

$$B_0 = \frac{1}{1 + i}A_0, \qquad B_1 = \lambda_1 V, \qquad B_2 = \frac{\lambda_2}{1 - i}A_2,$$

we want

$$F(B_1, B_1) = 1, \ F(B_0, B_2) = -i, \ F(B_2, B_0) = i, \ F(B_j, B_k) = 0 \text{ otherwise},$$

and

$$\det(B_0, B_1, B_2) = 1.$$

These reduce to the equations

(9) $$|\lambda_1|^2 F(V, V) = 1,$$

(10) $$\frac{\lambda_2}{(1 - i)^2} F(A_2, A_0) = i,$$

and

(11) $$\frac{\lambda_1 \lambda_2}{2} \det(A_0, A_1, A_2) = 1.$$

We use (10) and (11) to solve for λ_1 and λ_2. The solutions are clearly real. The equation (9) is automatic.

PROOF OF THEOREM 2. The anti-involution associated to P is given by

$$A(Z) = \sum \bar{z}_j A_j,$$

where

$$Z = \sum z_j A_j.$$

Thus

$$F(Z, Z) = z_k \bar{z}_\ell F(A_k, A_\ell)$$

and

$$\begin{aligned} F(A(Z), A(Z)) &= \bar{z}_k z_\ell F(A_k, A_\ell) \\ &= \bar{z}_\ell z_k F(A_\ell, A_k) \\ &= z_k \bar{z}_\ell \overline{F(A_k, A_\ell)}. \end{aligned}$$

Thus if each $F(A_k, A_\ell)$ is real, A preserves the quadratic form F and so maps $\tilde{S}$ to itself. Conversely, if A maps $\tilde{S}$ to itself, then $F(A_k, A_\ell) = \mu \overline{F(A_k, A_\ell)}$ for some μ. Take conjugates and interchange k and ℓ to see that μ must be $+1$ or -1. If $\mu = -1$ then $F(V, V)$ is imaginary, and hence zero, for each $V \in P$. But this cannot be. So $\mu = 1$, and $F(A_k, A_\ell)$ is real.

We now know which totally real planes project down to pseudocircles in Q (or S). The following exercise is easily proved directly without recourse to pseudocircles.

EXERCISE 6. Let P be a totally real 3-plane in $\mathbb{C}^3$ and assume $Q \cap [P]$ is a smooth curve. This curve is everywhere tangent to M if and only if $F(A_j, A_k)$ is real for each basis $\{A_1, A_2, A_3\}$ of P.

We may reformulate Theorem 1.

THEOREM 3. *The set of fixed points in Q of an anti-involution that preserves Q is a pseudocircle and all pseudocircles arise in this way.*

In the Euclidean plane three noncolinear points determine a circle. It is easy to see when three points on Q determine a pseudocircle.

THEOREM 4. *Let p, q, r be distinct points of $Q \subset \mathbb{C}^2$ that do not all lie on a complex line and let A_1, A_2, A_3 be vectors in $\mathbb{C}^3$ chosen so that $[A_1] = p$, $[A_2] = q$, $[A_3] = r$. The three points lie on a pseudocircle if and only if*

(12) $$F(A_1, A_2)F(A_2, A_3)F(A_3, A_1)$$

is real.

REMARK. This condition is independent of the particular vectors chosen to represent the given points and so is an intrinsic property of the points.

PROOF. If the points lie on a pseudocircle then the vectors can be chosen to lie on the fixed point set of the anti-involution and so each $F(A_j, A_k)$ is real. This of course is even stronger than (12). On the other hand, if (12) is true then it is possible to find complex numbers μ_1, μ_2, μ_3 so that for $B_j = \mu_j A_j$ we have that each $F(B_j, B_k)$ is real and so, since these vectors are also independent, they lie on the fixed point set of an anti-involution that maps $\tilde{Q}$ to itself. Hence they project down to points on a pseudocircle.

EXERCISE 7. Use this theorem to again prove that there is a three-parameter family of pseudocircles through any point of Q. (Hint: Show that there is a one-parameter family of pseudocircles through $(0, 0)$ and the point at infinity and therefore there is a one-parameter family through any two distinct points.)

§2. **The equations for pseudocircles.** We now determine the equations satisfied by pseudocircles. Our method is quite general. Let $M = G/H$ be a homogeneous space. Let $G_1 \in G$ be a subgroup and let $p \in M$ denote the coset eH. Let us assume that the orbit $G_1 p = \{g(p) : g \in G_1\}$ is a submanifold of M.

EXAMPLE. For $G = \mathbb{R}^2$ under vector addition and $H = \{(a, b) : a \text{ and } b \text{ are integers}\}$, G/H is a torus. If $G_1 = \{(c, d) : \frac{c}{d} = \omega\}$ where ω is a fixed number then $G_1 p$ is either a submanifold or is dense, depending on whether ω is rational or not.

Let Γ_0 denote the submanifold $G_1 p$ of M. We want to determine the equations satisfied by all the translates $\Gamma_g = g\Gamma_0$, $g \in G$. Each submanifold gG_1 of G projects onto the corresponding submanifold Γ_g of M. If we knew the equations giving gG_1, we could just project (i.e., eliminate fiber variables) to determine the equations of Γ_g. (Note that at this step it is best to have G_1 as large as possible so that we have to work with as few equations as possible.) But the equations of gG_1 are easily determined. G_1 is a subgroup of G. Its tangent space at the identity is given by the vanishing of some forms $\omega_1, \ldots, \omega_k$. These extend to be left-invariant forms on G (i.e., to be components of ω_{MC}). Then G_1 is an integral submanifold of $\{\omega_1, \ldots, \omega_k\}$ as is every left coset gG_1. This gives the equations for gG_1. Let us see how this actually works.

We start with the pseudocircle in Q, written in homogeneous coordinates,

(13) $$z_0 = (1 + i)u^2, \qquad z_1 = 2uv, \qquad z_2 = -(1 - i)v^2,$$

which is, as we have already seen, the fixed point set of the anti-involution

(14) $$A: (z_0, z_1, z_2) \rightarrow (i\bar{z}_0, \bar{z}_1, -i\bar{z}_2).$$

We repeatedly will make use of the observation that

$$z = \pm i\bar{z} \quad \text{if and only if} \quad z = (1 \pm i)\xi,$$

with ξ real.

We first determine those elements of $SU(2, 1)$ that map the fixed point set of A to itself (not necessarily leaving points fixed). We start with an element $B = (B_0, B_1, B_2)$ of $SU(2, 1)$ and write

$$B_j = \begin{pmatrix} B_{0j} \\ B_{1j} \\ B_{2j} \end{pmatrix}.$$

The action of B on any $q = (z_0, z_1, z_2) \in \mathbb{C}^3$ is

(15) $$Bq = z_0 B_1 + z_1 B_1 + z_2 B_2.$$

We require that

(16) $$A(Bq) = Bq,$$

for each fixed point q, i.e., for each q satisfying (13). We rewrite (13) as

(17) $$z_0 = i\bar{z}_0, \qquad z_1 = \bar{z}_1, \qquad z_2 = -i\bar{z}_2.$$

From (15), (16), and (17) we obtain

(18) $$\begin{cases} A(B_0) = iB_0, \\ A(B_1) = B_1, \\ A(B_2) = -iB_2. \end{cases}$$

That is,

$$i\overline{B_{00}} = iB_{00}, \qquad \overline{B_{10}} = iB_{10}, \qquad -i\overline{B_{20}} = iB_{20},$$

etc., which we write as

(19) $$\begin{cases} B_{00} = a_0, \ B_{10} = (1 - i)a_1, \ B_{20} = a_2 i, \\ B_{01} = (1 + i)b_0, \ B_{11} = b_1, \ B_{01} = (1 - i)b_2, \\ B_{02} = ic_0, \ B_{12} = (1 + i)c_1, \ B_{22} = c_2, \end{cases}$$

where all the new variables are real.

Recall that for B to be in $SU(2, 1)$ we require

$$F(B_1, B_1) = 1, \quad F(B_2, B_0) = i, \quad F(B_0, B_2) = -i, \quad \det(B) = 1,$$

and all other $F(B_j, B_k)$ are zero. Also recall that

(20) $$F(Z, W) = -iz_0\bar{w}_2 + iz_2\bar{w}_0 + z_1\bar{w}_1.$$

These lead to the equations

$$
(21) \quad
\begin{cases}
4b_0 b_2 + b_1^2 = 1, \\
c_0 a_2 + c_2 a_0 + 2c_1 a_1 = 1, \\
- a_0 a_2 + a_1^2 = 0, \\
c_0 c_2 + c_1^2 = 0, \\
a_0 b_2 - a_2 b_0 + a_1 b_1 = 0, \\
c_0 b_2 + c_2 b_0 + c_1 b_1 = 0, \\
\det B = 1.
\end{cases}
$$

These equations are not independent. For if they were, then B would depend on two parameters and this would contradict something that we already know. Namely, the dimension of the subgroup leaving (13) fixed is three (see the second corollary to Theorem 2.5). In fact, the determinant condition can be dropped. To see this write (18) in matrix form as

$$
A\overline{B} = (iB_0, B_1, -iB_2).
$$

So

$$
\det \overline{B} = \det B.
$$

Since B preserves the quadratic form F, $|\det B| = 1$ and so

$$
\det B = \pm 1.
$$

We work only with the $+1$ component and drop the last equation of (21). It is easy to see directly that the remaining equations define a three-dimensional submanifold of $SU(2, 1)$: for the gradients of the left-hand sides are linearly independent at $a_0 = b_1 = c_2 = 1$, and $a_j = b_k = c_m = 0$ otherwise. Thus the solution set is a submanifold near the identity and therefore by group translation everywhere. We would like to compute the restriction of our Maurer–Cartan forms to this submanifold in order to find which components vanish. This can be done in a straightforward but long manner. But the particular choice of A allows us (following Cartan) to shorten the computation considerably.

Recall from (5.25) that

$$
(22) \quad
\begin{cases}
\Omega \;\; = -iF(dB_0, B_0), \\
\Omega_1 = F(dB_0, B_1), \\
\Omega_2 = iF(dB_0, B_2) - F(dB_1, B_1), \\
\Omega_3 = -F(dB_2, B_1), \\
\Omega_4 = -iF(dB_2, B_2).
\end{cases}
$$

Let J denote the matrix

$$J = \begin{pmatrix} 0 & 0 & 1 \\ 0 & 1 & 0 \\ 1 & 0 & 0 \end{pmatrix}.$$

(We use this notation only here and shortly abandon it.) Note that $Z \cdot JW = JZ \cdot W$.

From (14) and (20) we have

$$F(Z, W) = Z \cdot JA(W),$$

and so also

$$\overline{F(Z, W)} = W \cdot JA(Z).$$

Our basic technique is to compare each Ω_j with its conjugate. For instance,

$$\begin{aligned} \Omega = -iF(dB_0, B_0) &= -idB_0 \cdot JA(B_0) \\ &= -idB_0 \cdot J(iB_0) \\ &= dB_0 \cdot JB_0, \end{aligned}$$

while

$$\begin{aligned} \overline{\Omega} = iF(B_0, dB_0) &= iB_0 \cdot JA(dB_0) \\ &= iB_0 \cdot J(idB_0) \\ &= -B_0 \cdot J(dB_0) \\ &= -JB_0 \cdot dB_0 \\ &= -dB_0 \cdot JB_0. \end{aligned}$$

But Ω is real and thus

$$\Omega = 0.$$

Note that we have used that $A(dB_0) = idB_0$. This follows from $A(B_0) = iB_0$ since A is a linear operator.

Let us apply the same technique to Ω_1.

$$\Omega_1 = F(dB_0, B_1) = dB_0 \cdot JA(B_1) = dB_0 \cdot JB_1$$

and

$$\overline{\Omega}_1 = F(B_1, dB_0) = B_1 \cdot JA(dB_0) = iB_1 \cdot JdB_0.$$

Thus

$$\Omega_1 = -i\overline{\Omega}_1.$$

Similarly

$$\Omega_2 = idB_0 \cdot JA(B_2) - dB_1 \cdot JA(B_1) = dB_0 \cdot JB_2 - dB_1 \cdot JB_1$$

and

$$\Omega_2 = -iB_2 \cdot JA(dB_0) - B_1 \cdot JA(dB_1) = B_2 \cdot JdB_0 - B_1 \cdot JdB_1,$$

and thus

$$\Omega_2 = \overline{\Omega}_2.$$

Also

$$\Omega_3 = -dB_2 \cdot JA(B_1) = -dB_2 \cdot JB_1$$

and

$$\overline{\Omega}_3 = -B_1 \cdot JA(dB_2) = iB_1 \cdot JdB_2,$$

so

$$\Omega_3 = i\overline{\Omega}_3.$$

Finally,

$$\Omega_4 = -idB_2 \cdot JA(B_2) = -dB_2 \cdot JB_2$$

and

$$\overline{\Omega}_4 = iB_2 \cdot JA(dB_2) = B_2 \cdot JdB_2 = JB_2 \cdot dB_2,$$

but Ω_4 is real. Thus

$$\Omega_4 = 0.$$

The forms $\mathscr{F} = \{\Omega, \Omega_1 + i\overline{\Omega}_1, \Omega_2 - \overline{\Omega}_2, \Omega_3 - i\overline{\Omega}_3, \Omega_4\}$ are either real or multiples of real forms. They are linearly independent and form a closed differential ideal. Thus they define a foliation of $SU(2, 1)$ by three-dimensional submanifolds. Let G_1 be the subgroup of $SU(2, 1)$ that maps the pseudo-circle (13) to itself. So $B \in G_1$ if and only if (18) holds. The leaves of the foliation defined by $\mathscr{F}$ are the left cosets of G_1. By our general remarks earlier, each leaf projects down to a curve in Q. In this way we clearly obtain all the pseudocircles.

EXERCISE 8. Use this method to derive the chain equations. (Hint: Modify the result of Exercise 2.5 to find the 4-dimensional group that has the u-axis as an orbit. Then use (22) to evaluate Ω_1 and Ω_3 on this subgroup.)

We now find the equations of the pseudocircles. For the hyperquadric

$$\operatorname{Im} w = \frac{1}{2}|z|^2$$

we take z and $u = \operatorname{Re} w$ as our coordinates. Then, as in Chapter 6,

$$(23) \qquad \begin{cases} \omega = du + \dfrac{1}{2}iz\,d\bar{z} - \dfrac{1}{2}i\bar{z}\,dz, \\[2mm] \omega_1 = dz, \\[2mm] \omega_2 = \omega_3 = \omega_4 = 0, \end{cases}$$

and

$$
(24) \quad
\begin{cases}
\Omega = |\lambda|^2 \omega, \\
\Omega_1 = \lambda(\omega_1 + \mu\omega), \\
\Omega_2 = \dfrac{d\lambda}{\lambda} + \omega_2 - 2i\overline{\mu}\omega_1 - i\mu\overline{\omega}_1 + (\rho - \tfrac{3}{2}i|\mu|^2)\omega, \\
\Omega_3 = \dfrac{1}{\lambda}\left[d\mu + \omega_3 + \mu\overline{\omega}_2 + \left(\rho + \tfrac{1}{2}i|\mu|^2\right)\omega_1 + i\mu^2\overline{\omega}_1 \right. \\
\qquad\qquad\qquad\qquad\qquad \left. + \mu\left(\rho + \tfrac{i}{2}|\mu|^2\right)\omega \right], \\
\Omega_4 = \dfrac{1}{|\lambda|^2}\left[d\rho + \tfrac{i}{2}(\mu d\overline{\mu} - \overline{\mu} d\mu) - i\overline{\mu}\left(\rho + \tfrac{1}{2}i|\mu|^2\right)\omega_1 \right. \\
\qquad\qquad\qquad \left. + i\mu\left(\rho - \tfrac{1}{2}i|\mu|^2\right)\overline{\omega}_1 + \left(\rho^2 + \tfrac{1}{4}|\mu|^4\right)\omega \right].
\end{cases}
$$

So

$$
(25) \quad
\begin{cases}
\Omega = |\lambda|^2 \omega, \\
\Omega_1 = \lambda(\omega_1 + \mu\omega), \\
\Omega_2 = \dfrac{d\lambda}{\lambda} - 2i\overline{\mu}\omega_1 - i\mu\overline{\omega}_1 + \left(\rho - \tfrac{3}{2}i|\mu|^2\right)\omega, \\
\Omega_3 = \dfrac{1}{\lambda}\left[d\mu + \left(\rho + \tfrac{1}{2}i|\mu|^2\right)\omega_1 + i\mu^2\overline{\omega}_1 + \mu\left(\rho + \tfrac{i}{2}|\mu|^2\right)\omega \right], \\
\Omega_4 = \dfrac{1}{|\lambda|^2}\left[d\rho + \tfrac{i}{2}(\mu d\overline{\mu} - \overline{\mu} d\mu) - i\overline{\mu}\left(\rho + \tfrac{1}{2}i|\mu|^2\right)\omega_1 \right. \\
\qquad\qquad\qquad \left. + i\mu\left(\rho - \tfrac{1}{2}i|\mu|^2\right)\overline{\omega}_1 + \left(\rho^2 + \tfrac{1}{4}|\mu|^4\right) \right].
\end{cases}
$$

Our equations are

$$
(26) \quad
\begin{cases}
\Omega = 0, \\
\Omega_1 + i\overline{\Omega}_1 = 0, \\
\Omega_2 - \overline{\Omega}_2 = 0, \\
\Omega_3 - i\overline{\Omega}_3 = 0, \\
\Omega_4 = 0.
\end{cases}
$$

We start with the above coordinates on Q and the fiber variables λ, μ, and ρ. Let us write

$$
\lambda = re^{i\theta},
$$

and introduce a new variable

$$
\alpha = ke^{i\theta}\mu,
$$

where

$$
k = (1 - i)/\sqrt{2}.
$$

Note that $|k| = 1$, $\bar{k} = ik$, and $k^2 = -i$. We shall see that r and $\mathrm{Re}(\alpha)$ are unrestrained and give the fiber variables for the foliation.

Since $\Omega = 0$, we set $\omega = 0$ in each equation in (25) and then substitute into (26). From $\Omega_1 + i\overline{\Omega}_1 = 0$, we see that

(27)
$$\omega_1 + ie^{-2i\theta}\overline{\omega}_1 = 0.$$

From $\Omega_2 - \overline{\Omega}_2 = 0$, we have

(28)
$$2d\theta - 3\bar{\alpha}ke^{i\theta}\omega_1 - 3\alpha\bar{k}e^{-i\theta}\overline{\omega}_1 = 0$$

and from $\Omega_3 - i\overline{\Omega}_3 = 0$,

(29)
$$\begin{cases} d\alpha - d\bar{\alpha} - i(\alpha + \bar{\alpha})d\theta \\ \quad + ke^{i\theta}\left(\rho + \dfrac{1}{2}i|\alpha|^2 - \bar{\alpha}^2k^2\right)\omega_1 \\ \quad - ike^{-i\theta}\left(\rho - \dfrac{1}{2}i|\alpha|^2 - \alpha^2\bar{k}^2\right)\overline{\omega}_1 = 0. \end{cases}$$

Finally, $\Omega_4 = 0$ yields

(30)
$$\begin{cases} d\rho + \dfrac{i}{2}(\alpha d\bar{\alpha} - \bar{\alpha}d\alpha) - |\alpha|^2 d\theta \\ \quad - i\bar{\alpha}ke^{i\theta}\left(\rho + \dfrac{1}{2}i|\alpha|^2\right)\omega_1 \\ \quad + i\alpha\bar{k}e^{-i\theta}\left(\rho - \dfrac{1}{2}i|\alpha|^2\right)\overline{\omega}_1 = 0. \end{cases}$$

We know from the derivation of the equations that the solution to (27)–(30) must project to a curve. That is, the integral submanifold intersected with the fiber is a two-dimensional surface. To see precisely which surface it is, we set $\omega_1 = 0$. This yields

$$d\theta = 0,$$
$$d\alpha - d\bar{\alpha} = 0,$$
$$d\rho + \frac{i}{2}(\alpha - \bar{\alpha})d\alpha = 0.$$

Let $\alpha = a + ib$; we see that b and θ are constants. Take as coordinates of the fiber r, θ, a, b, ρ. Then the surface is given by r and a arbitrary and

$$\theta = \theta_0, \qquad b = b_0, \qquad \rho = b_0 a + \rho_0.$$

Now we look at any curve in our eight-dimensional space satisfying (27)–(30) and transverse to these two-dimensional surfaces. The curve projects to a pseudocircle. The values of r and a can vary arbitrarily so let us look for that curve with $r = 1$ and $a = 0$. Also, we can parametrize the pseudocircle so that $|\omega_1| = 1$. Then (27) determines ω_1 as a function of θ. Indeed, allowing for the choice of orientation,

(31)
$$-\omega_1 = \pm e^{(\frac{3}{4}\pi - \theta)i} = \pm ke^{-i\theta}.$$

Our equations (28), (29), (30) become

$$
\begin{aligned}
\frac{d\theta}{dt} &= \mp 3b, \\
\frac{db}{dt} &= \pm \rho, \\
\frac{d\rho}{dt} &= \mp 2b^3,
\end{aligned}
$$

(32)

and so the pseudocircles in Q are given by starting with any solution $\theta(t)$ to

(33)
$$
\theta''' = -\frac{2}{9}(\theta')^3
$$

and then integrating the vector field V given by

(34)
$$
\omega(V) = 0, \qquad \omega_1(V) = \pm \frac{(1-i)}{\sqrt{2}} e^{-i\theta(t)}.
$$

Note that we are free to prescribe $\theta(0)$, $\theta'(0)$, and $\theta''(0)$. Since ω is zero, the value of $\theta(0)$ determines the initial direction in H of the pseudocircle. So corresponding to each direction in H there is a two-parameter family of pseudocircles. Hence through each point there is a three-parameter family of pseudocircles. Compare this with Exercises 1 and 7.

EXERCISE 9. The simplest solution to (33) is $\theta \equiv 0$. Verify that the family of pseudocircles given by (34) for this solution includes our original pseudocircle (4).

EXERCISE 10. The variable r clearly does not occur in the equations (28), (29), (30). The variable a does. We have indicated above that it is legitimate to set $a = 0$ when determining the equation for $\theta(t)$. The skeptic should leave a in the equations, work out the equations corresponding to (32), which will contain a, eliminate b and ρ, note that a is thereby automatically eliminated, and end up with (33).

Finally, we define pseudocircles on an arbitrary strictly pseudoconvex M^3. We have that the differential ideal for SU(2,1),

(35)
$$
\{\Omega, \Omega_1 + i\overline{\Omega}_1, \Omega_2 - \overline{\Omega}_2, \Omega_3 - i\overline{\Omega}_3, \Omega_4\},
$$

defines a foliation with three-dimensional leaves and that each leaf projects down to a curve on Q.

EXERCISE 11. (a) Let M^3 be a strictly pseudoconvex hypersurface. Recall the bundle B^8 over M and the forms $\Omega, \Omega_1, \ldots, \Omega_4$ that satisfy the equations (6.51). Show that the differential ideal (35) is also closed in this case.

(b) Show that each integral submanifold in B of (35) projects down to a curve in M.

Thus the following definition makes sense.

DEFINITION. A *pseudocircle* on M is the curve obtained by projecting into M any integral submanifold in B of the closed differential ideal

$$\{\Omega,\, \Omega_1 + i\overline{\Omega}_1,\, \Omega_2 - \overline{\Omega}_2,\, \Omega_3 - i\overline{\Omega}_3,\, \Omega_4\}.$$

Pseudocircles are even less well understood than chains.

CHAPTER 10

Nonsolvability of the Lewy Operator

§1. The basic results. Now we return to the questions of solvability which were briefly introduced in Chapter 1. We say that a partial differential operator L is solvable at a point P if for each function f (of an appropriate class) defined in a neighborhood of P there is some function u (of an appropriate class) defined on a possibly smaller neighborhood $\mathcal{O}$ such that

$$Lu = f \text{ on } \mathcal{O}.$$

Otherwise, we say that L is nonsolvable at P. We do not strive for the optimal result as far as these classes are concerned; but a good nonsolvability result would have f restricted and u general. So we take $f \in C^\infty$ and $u \in C^1$ (or $C^{1+\varepsilon}$). In fact, Theorem 1 is known to be true even if u is only a distribution [**Ho1**] or a hyperfunction [**Sc**]. Our first results are from [**JT1**] and [**JT2**] and generalize those of [**Le2**] and [**Ni2**].

The phenomenon of nonsolvability was first discovered by Lewy upon considering the induced Cauchy–Riemann operator on the hyperquadric. As we shall now see, the same result holds for any embedded strictly pseudoconvex hypersurface in $\mathbb{C}^2$.

All vector fields, choices for the Lewy operator, etc., are to be C^∞.

THEOREM 1. *Let $M^3 \subset \mathbb{C}^2$ be strictly pseudoconvex at the origin and let L be any choice for the Lewy operator. There exists a function $f \in C^\infty(M)$ with the property:*

> *there is no neighborhood $\mathcal{O}$ of the origin and function $u \in C^1(\mathcal{O})$ that satisfies $Lu = f$ in $\mathcal{O}$.*

PROOF. We may start with M in the form

$$v = |z|^2 + \alpha z u + \overline{\alpha}\overline{z}u + bu^2 + O(|z| + |u|)^3,$$

which we abbreviate as

(1) $$v = |z|^2 + G(z, \overline{z}, u).$$

A convenient choice for the Lewy operator will be

(2) $$L = (1 + iG_u)\frac{\partial}{\partial \overline{z}} - i(z + G_{\overline{z}})\frac{\partial}{\partial u}.$$

The set

$$\Gamma(\lambda) = \{(z, \lambda) \in \mathbb{C}^2 : \operatorname{Im} \lambda = |z|^2 + G(z, \overline{z}, \operatorname{Re} \lambda)\}$$

is the intersection of M with the complex line

$$\{(z, \lambda) : z \in \mathbb{C}, \lambda \text{ fixed }\}.$$

We write $\lambda = \xi + i\eta$ and also identify λ with $w = u + iv$.

EXERCISE 1. Show that near the origin in the λ-plane there is a curve γ given by $\eta = \eta(\xi)$ such that

(a) if $\operatorname{Im} \lambda > \eta(\operatorname{Re} \lambda)$, then $\Gamma(\lambda)$ is a simple closed curve,

(b) if $\operatorname{Im} \lambda = \eta(\operatorname{Re} \lambda)$, then $\Gamma(\lambda)$ is a point,

(c) if $\operatorname{Im} \lambda < \eta(\operatorname{Re} \lambda)$, then $\Gamma(\lambda)$ is empty.

(Hint: This exercise and the next are lemmas in [**JT1**].)

Let Ω be an open set, diffeomorphic to the ball, in the λ-plane that lies above γ and contains an arc of γ in its boundary. Let $\mathcal{O}$ be an open ball whose closure, $\overline{\mathcal{O}}$, is contained in Ω. Set

$$S = \{(z, \lambda) \in M : \lambda \in \operatorname{bdy} \mathcal{O}\}$$

and

$$T = \{(z, \lambda) \in M : \lambda \in \mathcal{O}\}.$$

So T is an embedded solid torus and S is the embedded torus that bounds T.

For any set $\mathbf{S}$ in the λ-plane, let

$$\Gamma(\mathbf{S}) = \{(z, \lambda) \in M : \lambda \in \mathbf{S}\}.$$

So $\Gamma(\operatorname{bdy} \mathcal{O}) = S$ and $\Gamma(\mathcal{O}) = T$.

We now study the line integral

$$I(\lambda) = \int_{\Gamma(\lambda)} h \, dz$$

as a complex-valued function of λ. The curve along which we integrate varies holomorphically (in an obvious sense) with λ. If h is a CR function then it is reasonable to guess that $I(\lambda)$ is a holomorphic function.

LEMMA 1. *Let h be a continuously differentiable CR function on some open set $\mathcal{D} \subset M$. Then $I(\lambda)$ is a holomorphic function of λ, as long as $\Gamma(\lambda) \subset \mathcal{D}$.*

PROOF. In a neighborhood of $\Gamma(\lambda)$ we may introduce coordinates $\lambda, \overline{\lambda}, \theta$. Then

$$I(\lambda) = \int_0^{2\pi} h z_\theta \, d\theta$$

and

$$\frac{\partial}{\partial \overline{\lambda}} \int_0^{2\pi} h \, dz = \int_0^{2\pi} \left(h_{\overline{\lambda}} z_\theta - h_\theta z_{\overline{\lambda}} \right) d\theta.$$

Now dz, $d\bar{z}$, and dw are linearly independent one-forms on M. Since L annihilates the functions z and w, while $L\bar{z} = 1 + iG_u$, we have

$$dh = \alpha dz + \beta d\bar{z} + \gamma dw$$

with $L(h) = \beta(1 + iG_u)$. Thus $\beta = 0$. The λ coordinate is the same as the w coordinate. Therefore we write

$$dh = \alpha dz + \gamma d\lambda.$$

But we also have

$$dh = h_{\bar{\lambda}} d\bar{\lambda} + h_\theta d\theta + h_\lambda d\lambda.$$

So $h_{\bar{\lambda}} = \alpha z_{\bar{\lambda}}$ and $h_\theta = \alpha z_\theta$. Thus $h_{\bar{\lambda}} z_\theta - h_\theta z_{\bar{\lambda}} = 0$ and we are done.

COROLLARY. *If h is a CR function on $\Gamma(\Omega) - T$, and is continuous on the closure of this set, then $I(\lambda) = 0$ for all $\lambda \in \Omega - \mathscr{O}$.*

PROOF. $I(\lambda)$ is holomorphic on $\Omega - \mathscr{O}$, is continuous to the boundary, and is zero on the arc, $(\text{bdy}\,\Omega) \cap \gamma$. Thus it is identically zero.

COROLLARY. *For h as above*

$$\iint_S h \, dz \, dw = 0.$$

PROOF. There is no trouble in writing the double integral as an iterated integral;

$$\iint_S h \, dz \, dw = \int_{\text{bdy}\,\mathscr{O}} I(\lambda) d\lambda = 0.$$

EXERCISE 2. Let h be any C^1 function in a neighborhood of an open set $\mathscr{D} \subset M^3$, where $\mathscr{D}$ has a smooth boundary. Let $z = x + iy$ and $w = u + iv(x, y, u)$. Then

$$\iint_{\text{bdy}\,\mathscr{D}} h \, dz \, dw = 2i \iiint_{\mathscr{D}} (Lh) \, dx \, dy \, du.$$

It is now an easy matter to prove Theorem 1. First we find an f for which

$$(3) \qquad\qquad\qquad Lu = f$$

is not solvable on a particular neighborhood of the origin. Then we use a standard trick to find an f for which (3) is unsolvable on each neighborhood, no matter how small. So fix open sets $\mathscr{O}$ and Ω as above and let $f \in C^\infty$ be positive in T and zero outside of T. We claim that (3) is unsolvable on each open set $\tilde{\Omega}$ containing $\Gamma(\Omega)$. For, on the contrary, assume there is some $u \in C^1(\tilde{\Omega})$ with $Lu = f$. From Exercise 2, we see that

$$(4) \qquad\qquad i\iint_S u \, dz \, dw = -2 \iiint_T f \, dx \, dy \, du < 0.$$

But from the second corollary above, we have that the left-hand side is zero.

To find an f that works for all neighborhoods of the origin, we pick a sequence of balls $\mathcal{O}_j$ with the properties:

    (i) $\mathcal{O}_j$ lies above γ,

    (ii) $\mathcal{O}_j$ and $\mathcal{O}_k$ are disjoint for $j \neq k$,

    (iii) $\lim_{j \to \infty} \mathcal{O}_j = \{0\}$,

    (iv) for any ball B centered at the origin

$$\mathcal{D} = \{B - \bigcup_j \mathcal{O}_j\} \cap \{(\xi, \eta) : \eta > \eta(\xi)\}$$

    is arc connected.

Note that a function holomorphic on $\mathcal{D}$ is identically zero if it is both continuous to the boundary and zero on some arc of the boundary.

Now take $f \in C^\infty$ to be positive in $\bigcup T_j$ and zero elsewhere. We claim that there is no open set $\mathcal{U}$ containing the origin and function $u \in C^1(\mathcal{U})$ with $Lu = f$ on $\mathcal{U}$. For if there were, then for some J and some corresponding Ω, we would have that $Lu = 0$ on $\Gamma(\Omega - \bigcup_J^\infty \mathcal{O}_j)$ and so

$$\iint_{S_j} u \, dz \, dw = 0, \qquad j \geq J.$$

But (4) holds for $S = S_j$ and $T = T_j$. So we have our contradiction.

For Theorem 1 we started with a realizable CR structure and showed that its Lewy operator is nonsolvable. Now we modify the construction to obtain nonrealizable structures.

THEOREM 2. *Let M be strictly pseudoconvex at the origin and let L be any choice for the Lewy operator. There exists some vector field L', which agrees with L to infinite order at the origin, with the property:*

    the only C^1 functions that satisfy $L'h = 0$ in a connected neighborhood of the origin are the constants.

REMARK. At each point of a realizable CR structure, at least one of the coordinate functions restricts to a CR function with nonzero differential. Thus the CR structure defined by L' is not realizable. Soon we will find structures that are even further from being realizable. Recall that in Chapter 1 we found a nonrealizable structure but one with a degenerate Lewy form. Also recall that Kuranishi proved that all strictly pseudoconvex structures of large enough dimension are realizable.

PROOF. We start with the sequence $\{\mathcal{O}_j\}$ used above and the associated solid tori, but now we take two functions f and g that satisfy

    (i) $f \in C^\infty$, f positive on $\bigcup_{j=0}^\infty T_{2j+1}$, f zero elsewhere,

    (ii) $g \in C^\infty$, g positive on $\bigcup_{j=1}^\infty T_{2j}$, g zero elsewhere.

Let L be the Lewy operator chosen in (2) and let

(5)
$$L' = L + f\frac{\partial}{\partial z} + g\frac{\partial}{\partial u}.$$

We claim that if

(6)
$$L'h = 0$$

in a neighborhood of the origin, then $dh = 0$ at the origin. To see this, first note that outside $\bigcup T_j$, and close enough to the origin, we have $Lh = 0$. So again

$$\iint_{S_j} h\, dz\, dw = 0,$$

and thus

$$\iiint_{T_j} Lh\, dx\, dy\, du = 0.$$

That is,

(7)
$$\iiint_{T_{2j+1}} -fh_z\, dx\, dy\, du = 0$$

and

(8)
$$\iiint_{T_{2j}} -gh_u\, dx\, dy\, du = 0.$$

From (7) we see that there are points p_{2j+1} and q_{2j+1} in T_{2j+1} at which

$$\operatorname{Re}h_z(p_{2j+1}) = 0, \quad \operatorname{Im}h_z(q_{2j+1}) = 0.$$

Thus $h_z(0) = 0$. In the same way, from (8) we obtain $h_u(0) = 0$. Finally, the equation (6) itself provides that $h_{\bar{z}}(0) = 0$. So we are done.

We now improve Theorem 2. First we show that almost for free we can eliminate the distinguished point and get the much stronger result that if $L'u = 0$ on some connected set then u is identically a constant on that set. The proof is a Baire category argument based on the one in [Le2]. A slight restriction on the smoothness of u is all that we must pay. By applying the Baire category argument to the same space with a finer topology, we can guarantee that L' is strictly pseudoconvex on all of $\mathbb{R}^3$. Then we show that a more complicated construction allows us to obtain the same results without this extra regularity. This argument is due to Fornæss [personal communication]. Essentially, it replaces the equalities in [JT2] by inequalities and the Baire category argument by a construction.

Let Ω be some open subset of $\mathbb{R}^3$. The usual topology on $C^\infty(\Omega, \mathbb{C})$ is defined by means of the seminorms

$$\rho_j(f) = \max_{\substack{x\in K_j \\ |\alpha|\leq j}} |D^\alpha f(x)|,$$

where $K_j \subset\subset K_{j+1}$ and $\bigcup K_j = \Omega$.

This topology is metrizable. In fact,

$$\rho(f, g) = \sum 2^{-j} \frac{\rho_j(f - g)}{1 + \rho_j(f - g)}$$

is a metric.

EXERCISE 3. (a) Show that $f_n \to 0$ in this metric if and only if $\lim_{n \to \infty} \rho_j(f_n)$
$= 0$ for each j. Note that this is very different from controlling $\lim_{j \to \infty} \rho_j(f_n)$—
convergence in $C^\infty(\Omega)$ does not imply control of f_n near the boundary of
Ω.

(b) Given any ε, fix some $\bar{n} > -\ln(\frac{1}{2}\varepsilon)$. Show that $\rho(f, g) < \varepsilon$ provided
$\sum^{\bar{n}} \rho_j(f - g) < \frac{1}{2}\varepsilon$.

We topologize the space of operators

$$\mathcal{S} = \left\{ \sum_1^3 \alpha_j(x) \frac{\partial}{\partial x_j} : \alpha_j \in C^\infty(\Omega, \mathbb{C}) \right\}$$

by using this C^∞ topology for each coefficient.

DEFINITION. An operator $L \in \mathcal{S}$ is called ε-*aberrant* if for every connected subset Ω_0 of Ω and every ε, $0 < \varepsilon < 1$, the only functions that
satisfy

$$(9) \qquad\qquad h \in C^{1+\varepsilon}(\Omega_0), \quad Lh = 0 \text{ on } \Omega_0,$$

are the constants.

THEOREM 3. *The ε-aberrant operators are dense in $\mathcal{S}$.*

The proof is surprisingly easy. Recall the Baire Category Theorem (see,
for instance, [**Roy**, p. 121]): a complete metric space cannot be written as a
countable union of nowhere dense sets. It follows that the countable union
of nowhere dense sets is, in fact, nowhere dense. Thus the complement of
the closure of the union is dense.

Let $\{P_j\}$ be a countable dense set of points in Ω, N_j the ball of radius
$\frac{1}{j}$ centered at P_j, $\|h\|_{\varepsilon, j}$ the Hölder norm for $C^{1+\varepsilon}(N_j)$, and $|w(P)|$ any
smoothly varying norm on the three-dimensional vector space of 1-forms at
P. Let E_j be the set of all those vector fields $L \in \mathcal{S}$ for which there is
some h with the properties:

$$(10) \qquad \begin{cases} Lh = 0 \quad \text{in } N_j, \\ \|h\|_{1+\frac{1}{j}, j} \leq j, \\ |dh(P_j)| \geq \frac{1}{j}. \end{cases}$$

LEMMA 2. *The set of ε-aberrant operators is the complement of $\bigcup E_j$.*

PROOF. Certainly, if L is in the union then L is not ε-aberrant. Conversely, if L is not ε-aberrant then for some connected Ω_0 and some ε, we
have a non-constant function $h \in C^{1+\varepsilon}(\Omega_0)$ with $Lh = 0$ on Ω_0. So there

exists some $P \in \Omega_0$ with $dh(P) \neq 0$. It follows that $L \in E_j$ if P_j is close enough to P and j is large enough.

So to prove the theorem we need only show that each E_j is nowhere dense. To do this, we show that the closure of E_j does not contain an open set. First we find an easily characterized set that contains this closure. So let L belong to the closure of some E_j. Write P for P_j, ε for $\frac{1}{j}$, etc. There exist operators $L_k \in E$ and functions $h_k \in C^{1+\varepsilon}(N)$ such that

$$L_k \longrightarrow L \text{ in } C^\infty(\Omega),$$
$$L_k h_k = 0 \text{ in } N,$$
$$\|h_k\|_{1+\varepsilon, N} \leq A,$$
$$|dh_k(P)| \geq \varepsilon.$$

It follows that some subsequence of $\{h_k\}$ converges in C^1. Call the limit h. Thus the closure of E_j is contained in the set of operators L for which there exists some $h \in C^1$ satisfying

(11) $\qquad\qquad Lh = 0 \text{ in } N \quad \text{and} \quad |dh(P)| \geq \varepsilon.$

Thus to show E is nowhere dense, it suffices to start with $L \in E$ and to find for each $\delta > 0$ an operator $L' \in \mathscr{S}$ such that $\rho(L - L') < \delta$ but there is no function h that satisfies (11) with L' in place of L.

We can certainly find some operator L_1 with $\rho(L-L_1) < \frac{1}{3}\delta$ such that L_1 is real analytic and strictly pseudoconvex near P and agrees with L outside some small neighborhood U of P. Next we can use Theorem 2 to find an operator L_2 that agrees with L_1 to infinite order at P and has the property that $L_2 h = 0$ in a neighborhood of P implies $dh(P) = 0$. Further, L_2 can be chosen to also be identical with L outside of U. Now, L_2 has the form

$$L_2 = L_1 + f\frac{\partial}{\partial z} + g\frac{\partial}{\partial u},$$

and we know from the proof of Theorem 2 that, for any nonzero constant η, the operator

$$L_3 = L_1 + \eta f\frac{\partial}{\partial z} + \eta g\frac{\partial}{\partial u}$$

has these same properties. We take $|\eta|$ small enough so that $\rho(L_1 - L_3) < \frac{1}{3}\delta$. So $\rho(L - L_3) < \delta$, but L_3 cannot be in the closure of E since any C^1 function h satisfying the first equation in (11) for L replaced by L_3 must have $dh(P) = 0$.

This proof that the ε-aberrant operators are dense does not quite show the existence of a nowhere realizable, strictly pseudoconvex CR structure on $\mathbb{R}^3$. This is due to the fact that a neighborhood in C^∞ does not provide control at points of $\mathbb{R}^3$ near infinity. So any C^∞ neighborhood of a strictly pseudoconvex structure contains structures that are not everywhere strictly pseudoconvex and, in particular, the ε-aberrant operators we have just obtained need not be everywhere strictly pseudoconvex. However, our perturbations are compactly supported and it is reasonable to expect we can overcome this lack of control at infinity. Our way of doing this will be to work with a different topology on C^∞.

For a noncompact manifold M, there is a topology on the function space $C^\infty(M)$ much finer than the usual topology. This is the Whitney topology. We describe it for $C^\infty(\mathbb{R}^3)$. See [GG] for the general case and for the details we omit. For each continuous positive function $\delta: \mathbb{R}^3 \to \mathbb{R}$, and each integer k, we set

$$B_{k,\delta} = \left\{ f \in C^\infty(\mathbb{R}^3): \max_{|\alpha| \le k} |D^\alpha f(x)| < \delta(x), \text{ for all } x \in \mathbb{R}^3 \right\}.$$

For brevity, we introduce the notation

$$|f|_k = \max_{\substack{|\alpha| \le k \\ x \in \mathbb{R}^3}} |D^\alpha f(x)|.$$

The Whitney topology is defined by taking all such $B_{k,\delta}$ to be a basis for the open sets that contain the origin. This topology has more open sets than the usual C^∞ topology and so convergence is more restricted.

LEMMA 4. $\{f_n\}$ converges to zero in the Whitney topology for $C^\infty(\mathbb{R}^3)$ if and only if there is some compact set K such that

  (i) f_n and all its derivatives converge to zero uniformly on K, and
  (ii) there is some N such that f_n is equal to zero on the complement of K for all $n > N$.

PROOF. Assume that (i) and (ii) hold. To show that $\{f_n\}$ converges to zero we need to show that for each open set $\mathcal{O}$ that contains zero, there is some N^* so that $f_n \in \mathcal{O}$ whenever $n > N^*$. Since $\{B_{k,\delta}\}$ is a basis, we can find some $\delta(x)$ and some k so that $B_{k,\delta} \subset \mathcal{O}$. There is some N_1 such that

$$|f_n|_{k,K} < \inf_K \delta(x),$$

for $n > N_1$. So if $n > N^* = \max(N_1, N)$,

$$|f_n|_k < \delta(x).$$

Thus $f_n \in B_{k,\delta} \subset \mathcal{O}$.

Conversely, let $\{f_n\}$ converge to zero. Thus, given any $\delta(x)$ and any k, there is some N_1 such that

$$|f_n|_k < \delta(x),$$

for $n > N_1$. Letting δ equal various small constants, we see that $\{f_n\}$ converges uniformly, with all its derivatives, on all of $\mathbb{R}^3$. So we only have to show (ii). If K did not exist, there would be a subsequence $\{f_{n_k}\}$, a sequence of points $\{x_k\}$ converging to infinity, and a sequence of positive numbers $\{\varepsilon_k\}$ such that

$$|f_{n_k}(x_k)| > \varepsilon_k.$$

Let $\delta(x)$ be any continuous function with $\delta(x_k) = \varepsilon_k$. The open set

$$B_{0,\delta} = \{f: |f(x)| < \delta(x)\}$$

is a neighborhood of zero but there are infinitely many elements in the sequence $\{f_n\}$ that do not belong to $B_{0,\delta}$. Thus $\{f_n\}$ does not converge to zero. This contradiction shows that K must exist.

EXERCISE 4. Show that the Whitney topology on $C^\infty(\mathbb{R}^3)$ does not satisfy the first axiom of countability. That is, show that there is no countable basis for the open sets containing a given function. { Hint: This is easy. See [GG, p. 44]. }

It follows that this topology is not metrizable. So it is surprising that the conclusion of the Baire Category Theorem still holds:

THEOREM [GG]. *The countable intersection of open dense sets is dense.*

It is easy to verify that our previous arguments work also in this new topology. Take $\{P_j\}$, $\{N_j\}$, and $\{E_j\}$ as before. Our open sets are the complements of the closures of $\{E_j\}$ (in the Whitney topology). So we need to show that these closures are nowhere dense. It is enough to show that the closures in the usual C^∞ topology do not contain any set open in the Whitney topology. Fix some j and set $E = E_j$, etc. So given any operator $L \in E$ and any positive function $\delta(x)$ and any integer k, we need to find some operator L' with

$$|L - L'|_k < \delta(x),$$

and for which there is no function $h \in C^1(N)$ satisfying

(12) $L'h = 0 \text{ in } N \quad \text{and} \quad |dh(P)| \geq \varepsilon.$

Fix some small neighborhood U of P in N and set $\delta = \inf_{x \in U} \delta(x)$. We take L_1, L_2, and L_3 as before and set $L' = L_3$. If L_1 and η are chosen appropriately, we end up with

$$|L - L'|_{k,U} < \delta,$$

which implies, since the operators agree outside of U,

$$|L - L'|_k < \delta(x).$$

From this it follows that the ε-aberrant operators are dense even in the Whitney topology. This allows us to find a nowhere realizable CR structure that is strictly pseudoconvex on all of $\mathbb{R}^3$:

THEOREM 4. *There exists some C^∞ vector field L defined on $\mathbb{R}^3$ that has the properties*:

   1) L, $\overline{L}$, *and* $[L, \overline{L}]$ *are linearly independent everywhere on* $\mathbb{R}^3$, *and*
   2) *if* $h \in C^{1+\varepsilon}(\Omega)$ *for some open set* Ω *and some* $\varepsilon > 0$ *and* $Lh = 0$ *on* Ω, *then* h *is a constant on the components of* Ω.

PROOF. We may start, for instance, with

$$L_0 = \frac{\partial}{\partial \overline{z}} - iz\frac{\partial}{\partial u}.$$

It is easy to find a function $\delta(x) = \delta(z, \overline{z}, u)$ with the property that

(13) $|L - L_0|_1 < \delta(x)$

implies that L satisfies 1). Indeed, a small constant suffices. But we now know that the open neighborhood of L_0 given by (13) contains an ε-aberrant operator.

§2. **An improvement.** J. Fornæss has pointed out that Theorem 3 is true even for the class $C^1(\Omega)$. We now present his idea which involves carefully redoing the basic construction to maintain more control. For convenience we work on all of $\mathbb{R}^3$. So now $\Omega = \mathbb{R}^3$ and

$$\mathscr{S} = \left\{ L = \sum_1^3 \alpha_j(x)\frac{\partial}{\partial x_j} : \alpha_j \in C^\infty(\mathbb{R}^3, \mathbb{C}) \right\}.$$

DEFINITION. $L \in \mathscr{S}$ is *aberrant* if for every connected subset $\Omega_0 \in \mathbb{R}^3$ the only functions that satisfy

(14) $h \in C^1(\Omega_0),   Lh = 0 \text{ in } \Omega_0,$

are the constants.

THEOREM 5. *The aberrant operators are dense in* $\mathscr{S}$.

We start with a C^∞ vector field L defined on $\mathbb{R}^3$ that is strictly pseudo-convex at some point p. We can introduce coordinates so that, near p, z and $u + i(|z|^2 + G(z, \overline{z}, u))$ are the CR functions that provide the realization into $\mathbb{C}^2$. We then replace L by the appropriate multiple to have again

(15) $L = (1 + iG_u)\frac{\partial}{\partial \overline{z}} - i(z + G_{\overline{z}})\frac{\partial}{\partial u}.$

Next we choose three sequences of open sets in $\mathbb{C}$, the λ-plane, namely, $\{\mathscr{O}_i\}$, $\{\tilde{\mathscr{O}}_i\}$, and $\{\Omega_i\}$, with the following properties:

   (a) each set is diffeomorphic to the ball and has a smooth boundary whose length goes to zero,
   (b) $\mathscr{O}_i \subset\subset \tilde{\mathscr{O}}_i \subset\subset \Omega_i$,
   (c) Ω_i lies above γ and bdy Ω_i contains an arc of γ,
   (d) $\{\overline{\Omega}_i\}$ are disjoint,
   (e) Ω_i converges to the origin.

In addition, we assume these sequences start with small enough sets so that certain geometric quantities such as the area of Ω_j are all less than 1.

Again, for any set $\mathbf{S}$ in the λ-plane let

$$\Gamma(\mathbf{S}) = \{(z, \lambda): \operatorname{Im} \lambda = |z|^2 + G(z, \bar{z}, \operatorname{Re} \lambda) \text{ and } \lambda \in \mathbf{S}\}.$$

Set

$$T_i = \Gamma(\mathcal{O}_i) \quad \text{and} \quad K_i = \Gamma(\overline{\Omega}_i).$$

Take two functions f and g that satisfy

(i) $f \in C^\infty$, f positive on $\bigcup_{j=0}^{\infty} T_{2j+1}$, f zero elsewhere,

(ii) $g \in C^\infty$, g positive on $\bigcup_{j=1}^{\infty} T_{2j}$, g zero elsewhere.

Let

$$L' = L + f\frac{\partial}{\partial z} + g\frac{\partial}{\partial u}.$$

Here is a strengthened version of Theorem 2. We are letting ψ_j equal f for j odd and equal g for j even. The functions $b_j(\sigma)$ are chosen in the proof. They are positive for $\sigma > 0$ and go to zero as σ approaches zero.

THEOREM 6. *Let $\{\varepsilon_j\}$ be a sequence of positive numbers that satisfy $\varepsilon_j + b_j(\varepsilon_j) < \iiint_{T_j} \psi_j$. Let S be any C^∞ vector field with $|S - L'| < \varepsilon_j$ on K_j for each j. If h is a C^1 function with $Sh = 0$ in some neighborhood of the distinguished point p, then $dh = 0$ at p.*

Note, in particular, that $L'h = 0$ implies $dh = 0$ at p. By showing this same result for certain perturbations S of L' we will be able to eventually eliminate the reference to the distinguished point.

We have already used the norm $|f|_n$ where some point x is understood. When n is zero we just write $|f|$. Given some set K, we write

$$\|f\|_K = \sup_K |f|,$$

$$\|f\|_{n,K} = \sup_K |f|_n,$$

$$\|f\| = \sup |f|, \quad \text{taken over either } \mathbb{C} \text{ or } \mathbb{R}^3.$$

The proof of Theorem 6 depends on two general facts about functions of a complex variable.

FACT I. Let g be a bounded measurable function on $\mathbb{C}$ with compact support. Let

$$H(z) = \frac{1}{2\pi i} \iint_{\mathbb{C}} \frac{g(\zeta)}{z - \zeta} d\zeta d\bar{\zeta}.$$

Then H is a continuous function, its distributional derivative satisfies

$$\frac{\partial H}{\partial \bar{z}} = g,$$

and

$$\|H\| \le C\|g\|,$$

where C depends only on the support of g.

EXERCISE 5. Prove this in the following manner.

1. The bound is valid since $\frac{1}{\zeta}$ is integrable on $\mathbb{C}$. Show continuity by estimating $H(z_1) - H(z_2)$. Your proof will probably yield the stronger result that H is Lipschitz. In fact, if g satisfies any Hölder condition, then H is continuously differentiable [**Fe**, p. 544].

2. Use the identity

$$\iint \frac{\psi_{\bar{z}}}{z - \zeta} \, d\zeta \, d\bar{\zeta} = 2\pi i \psi(z),$$

for smooth compactly supported functions to show $\frac{\partial H}{\partial \bar{z}} = g$ when g is smooth [**Ho3**, p. 3].

3. The statement about the distributional derivative is equivalent to showing

$$\iint H\phi_{\bar{z}} \, dz \, d\bar{z} = - \iint g\phi \, dz \, d\bar{z},$$

for all $\phi \in C_0^\infty(\mathbb{C})$. Do this by integrating $\frac{g(\zeta)\phi(z)}{z-\zeta}$ over $\mathbb{C}^2$ by means of iterated integrals.

FACT II. Let $\mathcal{U}$ be an open set in $\mathbb{C}$ with smooth boundary. Let Γ be a subarc of bdy $\mathcal{U}$ and let K be a compact subset in $\mathcal{U}$. There is a function $b_0(\Gamma, K, \delta)$ with $\lim_{\delta \to 0} b_0(\Gamma, K, \delta) = 0$ such that

$$\|F\|_K < b_0(\Gamma, K, \delta),$$

whenever F is holomorphic in $\mathcal{U}$ and continuous to the boundary, with $|F| \le 1$ on $\mathcal{U}$ and $|F| < \delta$ on Γ.

EXERCISE 6. Use the following outline to prove this fact.

1. $f = \ln|F|$ is subharmonic in $\mathcal{U}$ and thus $f \le u$ in Ω if $f \le u$ on bdy Ω and $\Delta u = 0$. (This essentially is a Hormander, op. cit., pp. 16–18.)

2. The Dirichlet problem $\Delta u = 0$ in $\mathcal{U}$, $u = s$ on bdy $\mathcal{U}$ has a solution for any function s continuous on bdy $\mathcal{U}$. (See [**CH**, pp. 306–312] for a proof that uses subharmonic functions. Note that $\mathcal{U}$ need not be simply-connected.)

3. Let $\Gamma \subset$ bdy $\mathcal{U}$ be the given arc. Let $\Gamma_1 \subset \Gamma$ be a subarc and let s be a continuous function on bdy $\mathcal{U}$ with $s = -1$ on Γ_1, $-1 \le s \le 0$ on Γ and $s = 0$ outside Γ. Let u solve the above Dirichlet problem. By the maximum principle for harmonic functions, $u < 0$ in $\mathcal{U}$.

4. It follows that $f < |\ln \delta| u$ in $\mathcal{U}$ and so, for any compact set $K \subset \mathcal{U}$,

$$\|F\|_K < \delta^c \quad \text{where } c = -\max_K u > 0.$$

We start the proof of Theorem 6 with a generalization of Lemma 1.

LEMMA 5. *Let S be a complex vector field and h a C^1 function with $Sh = 0$ and $|dh| < 1$ on $K_j - T_j$. Let $\lambda \in \Omega_j - \overline{\mathscr{O}}_j$. If $|S - L'| < \sigma$ at each point of $\Gamma(\lambda)$ then*

$$\left| \frac{\partial}{\partial \overline{\lambda}} \int_{\Gamma(\lambda)} h \, dz \right| < C\sigma \,,$$

where C can be taken to be independent of j.

PROOF. As in the proof of Lemma 1, we introduce coordinates λ, $\overline{\lambda}$, θ and obtain that

$$\frac{\partial}{\partial \overline{\lambda}} \int_{\Gamma(\lambda)} h \, dz = \int_0^{2\pi} (h_{\overline{\lambda}} z_\theta - h_\theta z_{\overline{\lambda}}) \, d\theta \,.$$

Again we have

$$dh = \alpha \, dz + \beta \, d\overline{z} + \alpha \, dw \,.$$

We may determine β from

$$Lh = \beta (1 + i G_u) \,.$$

(For Lemma 1, we could conclude from this that $\beta = 0$.) From

$$dh = h_{\overline{\lambda}} d\overline{\lambda} + h_\theta d\theta + h_\lambda d\lambda \,,$$

we conclude

$$h_{\overline{\lambda}} = \alpha z_{\overline{\lambda}} + \beta \overline{z}_{\overline{\lambda}} \quad \text{and} \quad h_\theta = \alpha z_\theta + \beta \overline{z}_\theta \,.$$

Thus

$$h_{\overline{\lambda}} z_\theta - h_\theta z_{\overline{\lambda}} = \beta (\overline{z}_{\overline{\lambda}} z_\theta - \overline{z}_\theta z_{\overline{\lambda}})$$

and

$$\left| \frac{\partial}{\partial \overline{\lambda}} \int_{\Gamma(\lambda)} h \, dz \right| < C_1 |\beta| < C \|Lh\|_{\Gamma(\lambda)} \,.$$

But in $K_j - T_j$ we have $L = L'$. So $Lh = L'h = (L' - S)h + Sh$. Thus under the hypotheses of the lemma $\|Lh\|_{\Gamma(\lambda)} < \sigma$. This concludes the proof.

LEMMA 6. *There exist functions $b_j(\sigma)$ with $\lim_{\sigma \to 0} b_j(\sigma) = 0$ such that*

$$\left| \int_{\Gamma(\lambda)} h \, dz \right| < b_j(\sigma), \quad \lambda \in \mathrm{bdy}\, \widetilde{\mathscr{O}}_j \,,$$

provided

(a) $Sh = 0$ on $K_j - T_j$,

(b) $\|S - L'\|_{K_j - T_j} < \sigma$, $\|h\|_{K_j - T_j} < 1$, *and* $\|dh\|_{K_j - T_j} < 1$.

PROOF. Set $I(\lambda) = \int_{\Gamma(\lambda)} h \, dz$. We assume that the curve $\Gamma(\lambda)$ is short for $\lambda \in \overline{\Omega}_j$ and so $\|I\|_{\Omega_j - \mathscr{O}_j} < 1$. From Lemma 5 we also have $\|\frac{\partial}{\partial \overline{\lambda}} I\|_{\Omega_j - \mathscr{O}_j} < C\sigma$.

Let us suppress the index j. Clearly I is continuous in $\overline{\Omega - \mathcal{O}}$ with $|I| \leq 1$ there. By Fact I there exists a function H with $\|H\| < C\sigma$ and $\frac{\partial H}{\partial \lambda} = \frac{\partial I}{\partial \lambda}$ in $\Omega - \mathcal{O}$. Let $G = H - I$. Thus G is holomorphic and satisfies $|G| < 1 + C\sigma$ in $\Omega - \mathcal{O}$. Also, since $I = 0$ on $\overline{\Omega} \cap \gamma$, we have

$$|G|_{\overline{\Omega} \cap \gamma} < C\sigma.$$

Now let $F = \frac{G}{C\sigma + 1}$. So F is holomorphic in $\Omega - \mathcal{O}$, continuous in the closure and satisfies $|F| < 1$; on $\overline{\Omega} \cap \gamma$, F satisfies $|F| < C\sigma$. Set $\delta = C\sigma$, $\Gamma = \overline{\Omega} \cap \gamma$, and $K = \mathrm{bdy}\,\widetilde{\mathcal{O}}$. Apply Fact II to obtain $\|F\|_K \leq b_0(\delta)$. So on $\mathrm{bdy}\,\widetilde{\mathcal{O}}$,

$$|I| \leq |I - H| + |H| \leq |G| + C\sigma \leq (1 + C\sigma)b_0(\delta) + C\sigma = b(\sigma).$$

Note that the function $b(\sigma)$ depends only on Ω, $\widetilde{\mathcal{O}}$, and $\mathcal{O}$ and goes to zero as σ goes to zero. This proves Lemma 6.

Let us use this to find an upper bound for $\iint h\,dz\,dw$. Let $\mathscr{C}$ be some curve in the λ-plane lying above γ. For any function h,

$$\iint_{\Gamma(\mathscr{C})} h\,dz\,dw = \int_{\mathscr{C}} \left(\int_{\Gamma(\lambda)} h\,dz \right) d\lambda,$$

and so

$$\left| \iint_{\Gamma(\mathscr{C})} h\,dz\,dw \right| \leq \max_{\lambda \in \mathscr{C}} \left| \int_{\Gamma(\lambda)} h\,dz \right|,$$

provided $\mathscr{C}$ has small length. Thus Lemma 6 provides the following upper bound for $|\iint h\,dz\,dw|$.

LEMMA 7. *Let $\widetilde{T}_j$ be the image of $\widetilde{\mathcal{O}}_j$ in M. Under the hypotheses of the previous lemma*

$$\left| \iint_{\mathrm{bdy}\,\widetilde{T}_j} h\,dz\,dw \right| < b_j(\sigma).$$

We now find a lower bound for $|\iint h\,dz\,dw|$. Recall the expression for L' in terms of L and $\{\psi_j\}$. We will work in some Ω_j. If j is odd let h' denote h_z while if j is even let h' denote h_u. Also recall that T_j is the image of $\mathcal{O}_j$ in $\mathbb{R}^3$.

LEMMA 8. *Assume*

1. $|S - L'| < \sigma$ on $\widetilde{T}_j$,
2. $Sh = 0$ on $\widetilde{T}_j$,
3. $\mathrm{Re}\,h' > \frac{1}{2}$ on T_j and $|dh| < 1$ on $\widetilde{T}_j$.

Then

$$\left| \iint_{\text{bdy } \tilde{T}_j} h \, dz \, dw \right| \geq \left(\iiint_{T_j} \psi_j \right) - 2\sigma.$$

PROOF. We start with the result of Exercise 2:

$$\iint_{\text{bdy } \mathscr{D}} h \, dz \, dw = 2i \iiint_{\mathscr{D}} (Lh) \, dx \, dy \, du.$$

So we have

$$\left| \iint_{\text{bdy } \tilde{T}_j} h \, dz \, dw \right| = \left| 2i \iiint_{\tilde{T}_j} Lh \, dx \, dy \, du \right|$$

$$= 2 \left| \iiint (L - L')h + \iiint L'h \right|$$

$$\geq 2 \left| \iiint_{T_j} \psi_j h' \right| - 2 \left| \iiint_{\tilde{T}_j} (L' - S)h \right|.$$

We have used that $(L' - L)h = \psi_j h'$ on K_j, $\text{supp } \psi_j = T_j$, and $Sh = 0$ on $\tilde{T}_j$. Thus

$$\left| \iint_{\text{bdy } \tilde{T}_j} h \, dz \, dw \right| \geq 2(1/2) \iiint_{T_j} \psi_j - 2\sigma \sup_{\tilde{T}_j} |dh|$$

$$\geq \iiint_{T_j} \psi_j - 2\sigma.$$

(We also used that $\tilde{T}_j$ has volume less than one.)

We are ready to prove Theorem 6. We proceed by contradiction. So let us assume that $dh \neq 0$ at p. We have that $L = (1 + iG_u)\frac{\partial}{\partial \bar{z}} - i(z + G_{\bar{z}})\frac{\partial}{\partial u}$ with $dG(p) = 0$ and that S agrees with L at the origin. So $Sh = 0$ implies $h_{\bar{z}}(p) = 0$. Thus at least one of $h_u(p)$ and $h_z(p)$ is nonzero. Let j be even if $|h_u(p)| \geq |h_z(p)|$. Otherwise take it odd.

We may add a constant to h and then multiply by another constant and still have a solution to $Sh = 0$. Thus we may assume that $h(p) = 0$ and $h'(p) = \frac{3}{4}$. Take j large enough so that on K_j we have $Sh = 0$, $|h| < 1$, $|dh| < 1$, and $\text{Re } h' > 1/2$. The hypotheses of Lemmas 7 and 8 are valid for $\sigma = \varepsilon_j$. Thus the bounds for $|\iint h \, dz \, dw|$ yield

$$-2\varepsilon_j + \iiint_{T_j} \psi_j < b_j(\varepsilon_j).$$

That is,

$$\iiint_{T_j} \psi_j < b_j(\varepsilon_j) + 2\varepsilon_j.$$

But ε_j was chosen so that the right-hand side is strictly less than the left-hand side. This contradiction shows that the assumption $dh(p) \neq 0$ cannot hold. Thus the theorem is proven.

Let us restate this theorem in a form that will be useful later. First we want to start with an arbitrary operator L_0. Fix some point p and some neighborhood $\mathscr{U}$. Given any δ and any n, there clearly is some L such that

1. L is real analytic and strictly pseudoconvex near p,
2. $L = L_0$ outside $\mathscr{U}$,
3. $\|L - L_0\|_n < \frac{\delta}{2}$.

As we have seen in Chapter 1, $Lh = 0$ has two independent solutions near p. So we proceed as before and take

$$L' = L + f\frac{\partial}{\partial z} + g\frac{\partial}{\partial u}.$$

We can multiply f and g by the same small constant in order to obtain

$$\|L' - L\|_n \leq \frac{\delta}{2}.$$

The following result now follows easily from Theorem 6.

THEOREM 7. *Let L_0 be any C^∞ complex vector field on $\mathbb{R}^3$ and p any point in $\mathbb{R}^3$. Given a neighborhood $\mathscr{U}$ of p, an integer n, and a number $\delta > 0$, there exist*

(a) *a C^∞ complex vector field L' that agrees with L_0 outside of $\mathscr{U}$ and satisfies*

$$|L_0 - L'|_n < \delta \quad in \ \mathscr{U},$$

(b) *a sequence $\{\varepsilon_j\}$ of positive numbers converging to zero,*

and

(c) *a sequence $\{K_j\}$ of compact sets converging to p, such that, for any C^∞ complex vector field S, if $\|S - L'\|_{K_j} < \varepsilon_j$ for each j and $Sh = 0$ in a neighborhood of p then $dh = 0$ at p.*

Now we can prove that the aberrant operators are dense (Theorem 5). We start with some L_0 and some $\varepsilon > 0$. We fix some $\bar{n}, \bar{n} > -\ln(\frac{1}{2}\varepsilon)$. All we need to do is find some L that is aberrant and satisfies $\|L - L_0\|_{\bar{n}} < \frac{1}{2}\varepsilon$.

Let $p_1, p_2, p_3, \ldots$ be a dense set of points in $\mathbb{R}^3$. We will construct for each n a complex vector field L_n, a sequence of real numbers $\varepsilon_{n,1}, \varepsilon_{n,2}, \ldots$ and a sequence of compact sets $K_{n,1}, K_{n,2}, \ldots$ with the properties

(i) $\lim_{j \to \infty} \varepsilon_{n,j} = 0$ and $\lim_{j \to \infty} K_{n,j} = p_n$,

(ii) $\|L_{n+1} - L_n\|_{n+\bar{n}} < \varepsilon 2^{-(n+2)}$, for $n \geq 0$,

(iii) $\|L_{n+1} - L_n\|_{K_{\nu,j}} < 2^{-n} \varepsilon_{\nu,j}$, for $1 \leq \nu \leq n$ and $j \geq 1$.

(iv) For each S satisfying $\|S - L_{n+1}\|_{K_{n+1,j}} < \varepsilon_{n+1,j}$ for some $n \geq 0$ and all $j \geq 1$, and for any C^1 function h, $Sh = 0$ in a neighborhood of p_{n+1} implies $dh = 0$ at p_{n+1}.

EXERCISE 7. Show that (ii) implies that $\{L_n\}$ converges in C^∞ to some L and $\|L - L_0\|_{\bar{n}} \leq \frac{1}{2}\varepsilon$. Thus also, by our choice of $\bar{n}$, $\|L - L_0\|_\infty < \varepsilon$.

It is easy to see that this limit L is aberrant. For let $Lh = 0$ on some open set. Thus $Lh = 0$ in a neighborhood of each p_N in that open set. From (iii) we have

$$\|L - L_N\|_{k_{N,j}} \leq \sum_{n=N}^\infty \|L_{n+1} - L_n\|_{k_{N,j}}$$

$$< \sum_{n=N}^\infty 2^{-n} \varepsilon_{N,j}$$

$$\leq \varepsilon_{N,j}.$$

So we may set $L = S$ in (iv). Thus $dh = 0$ at p_N for each p_N in the open set. Since $\{p_n\}$ is dense in $\mathbb{R}^3$ we have that $dh = 0$ at a dense set of points in the given open set and, as long as this set is connected, h is a constant.

So we start with L_0 and the point p_1. Let $\mathscr{U}$ be an arbitrary neighborhood of p_1. We apply Theorem 7 with $n = \bar{n}$ and $\delta = \varepsilon 2^{-2}$ to obtain $L_1 = L'$, and sequences $\{\varepsilon_{1,j}\}$ and $\{K_{ij}\}$ that satisfy conditions (i), (ii), (iv) for $n = 0$ while (iii) is vacuous.

Next assume we have already found $L_0, L_1, \ldots, L_N$ such that (i) through (iv) hold for $0 \leq n \leq N - 1$. Pick some neighborhood $\mathscr{U}$ of p_{N+1} that does not include any of the previous points p_m, $m \leq N$. Again we apply Theorem 7; this time for L_N, $\mathscr{U}$, p_{N+1} and $n = \bar{n} + N$, $\delta = \varepsilon_1 \varepsilon 2^{-(N+2)}$. We specify ε_1 below. For now just take $\varepsilon_1 < 1$. We obtain an operator $L_{N+1} = L'$ and sequences $\{\varepsilon_{N+1,j}\}$ and $\{K_{N+1,j}\}$ that satisfy (i), (ii), and (iv) for $n = N$. Now let us show (iii) also holds. Note that because $\mathscr{U}$ does not include any previous p_m we have that only finitely many of the sets $\mathscr{U} \cap K_{\nu,j}$, $1 \leq \nu \leq N$, $1 \leq j < \infty$, are nonempty. Take

$$\varepsilon_1 = \min\{1/2, \min\{\varepsilon_{\nu,j} \mid \mathscr{U} \cap K_{\nu,j} \neq \phi\}\}.$$

Thus when $\mathscr{U} \cap K_{\nu,j} = \phi$ we have $L_{N+1} = L_N$ on $K_{\nu,j}$ while on the finitely many other sets

$$\|L_{N+1} - L_N\|_{K_{\nu,j}} \leq \|L_{N+1} - L_N\|_{n+N} \leq \varepsilon_1 \varepsilon 2^{-(N+2)} < 2^{-N} \varepsilon_{\nu,j}.$$

This concludes the proof of the denseness of aberrant operators.

EXERCISE 8. Let $\delta(x)$ be any continuous positive function on $\mathbb{R}^3$. Use the above construction to show that for each $L_0 \in \mathscr{S}$ there is some aberrant operator L that satisfies

$$|L - L_0|_1 < \delta(x).$$

Conclude that there is a strictly pseudoconvex CR structure on $\mathbb{R}^3$ with the property that the only local continuously differentiable CR functions are the constants (cf. Theorem 4).

Notes

Chapter 1

§2. Cartan called his geometry "pseudoconformal" in order to emphasize that what he was studying was a generalization of the theory of one complex variable (conformal geometry). The term "CR manifold" was first used in [Gr]. A careful discussion of the basic definitions and elementary properties may be found in [Ta] where the CR structures are not restricted to be of hypersurface type. There is little overlap between this book and ours; it focuses on a problem not considered here, namely, the extension problem for CR functions, and is a good introduction to this field. Everyone should know the two basic extension results: A CR function defined on the boundary of a bounded open set with connected complement extends to a function holomorphic on the open set (Bochner-Hartog Theorem) and a CR function on a strictly pseudoconvex hypersurface extends as a holomorphic function to some open set that contains the hypersurface in its boundary (Lewy Theorem).

The proof that a real analytic almost complex structure is complex can be found, for instance, in [KN, volume 2]. The original Newlander-Nirenberg Theorem required smoothness of class $C^{2n+\lambda}$ for manifolds of dimension $2n$. This was improved to $C^{1+\lambda}$ in [NW]. There are now several very different proofs of the Newlander-Nirenberg Theorem; for instance, those of Malgrange [Ma], [Ni2], Kohn [FK], and Webster [We2]. The one-dimensional version of the Newlander-Nirenberg Theorem is the existence of conformal coordinates. The standard references are [Be] and [Che].

§3. Theorem 1 holds in a formal sense for C^∞ functions and hypersurfaces. That is, if $f = f(z, \overline{z}, u)$ is CR on

$$\operatorname{Im} W = v(z, \overline{z}, u)$$

then there exists a formal power series

$$F = \sum a_{lk} z^l w^k$$

such that the Taylor coefficients at the origin for $f(z, \overline{z}, u)$ come from the formal power series for $F(z, u + iv(z, \overline{z}, u))$. This can be proved by adapting to formal power series the proof of Theorem 1.

§4. There are several proofs of Theorem 2 in the literature. The earliest seems to be [AH] where, however, it is remarked, "Although we were unable to find a proof... in the literature, (the) theorem seems to have been known for a long time." The present author has known each of the proofs given in Chapter One long enough to be unsure of their origins, but he believes that he learned the first from H. Rossi and the second from C. LeBrun. Realizability results in the absence of analyticity involve subtle questions of partial differential equations. Some of these can be avoided if the CR structure is part of a compact CR manifold: A compact strictly pseudoconvex CR manifold of dimension greater than three is locally realizable [BdM]. The restriction to dimensions greater than three is essential. The nonrealizable structures of dimension three constructed in Chapter Ten can, in a trivial manner, be extended to strictly pseudoconvex structures on the sphere.

However, a compact CR manifold may be realizable in the neighborhood of each of its points without being globally realizable. There are two interesting and relatively simple examples of such structures on S^3. We have seen that

$$L = w \frac{\partial}{\partial \overline{z}} - z \frac{\partial}{\partial \overline{w}}$$

globally defines the CR structure on $S^3 \subset \mathbb{C}^2$. For $0 \leq t < 1$, the operator $L + t\overline{L}$ also defines a CR structure on S^3. This structure is real analytic, therefore, locally realizable. Indeed, there is a global immersion $S^3 \subset \mathbb{C}^3$ that realizes this structure. But for $t \neq 0$ there is no embedding: any function $f(z, \overline{z}, w, \overline{w})$ on S^3 that is annihilated by $L + t\overline{L}$ for some t, $0 < t < 1$, is even, $f(z, \overline{z}, w, \overline{w}) = f(-z, -\overline{z}, -w, -\overline{w})$. This result is due to [Bu] who adapted an example in [Ros1]. For a new proof and generalizations see [Fal].

This example is strictly pseudoconvex at every point of the sphere. Our second example is strictly pseudoconvex at no point. A CR structure is called Levi flat if the Levi form is identically zero. It is easy to see that any Levi flat CR structure is locally realizable. On the other hand, any compact hypersurface in a Euclidean space has a point at which it is strictly convex. So it follows from the proof of Lemma 4 that any compact hypersurface in $\mathbb{C}^2$ has a point at which it is strictly pseudoconvex. Thus it only remains to construct a Levi flat CR structure on S^3 in order to have a locally, but not globally, realizable structure. A Levi flat structure provides a foliation of S^3 with two-dimensional leaves. So we start with one such foliation. See, for instance, [La] for a discussion of the Reeb foliation. This foliation uses the decomposition of S^3 into two solid tori joined along their common torus boundary. We use this common boundary to define an orientation on each leaf. We then place any metric on S^3 and let the J operator on any leaf be rotation by ninety degrees. This gives us the desired Levi flat CR structure.

Note that this structure induces a complex structure on each leaf and also that the foliation has a compact leaf, namely, the common torus boundary. So any global CR map into any $\mathbb{C}^n$ must map this torus to a point (and indeed any global CR function must be a constant on all of S^3). This is a second reason why the CR structure is not globally realizable.

§**6.** There are even strictly pseudoconvex CR structures, with the symmetry of Reinhardt hypersurfaces, which have the constants as the only global CR functions [**Ba**]. In Chapter Ten, we will find a CR structure on which the constants are even the only local CR functions. Such a structure, of course, is not locally realizable at any of its points.

Chapter 3

The basic technique in this chapter is to equate coefficients in Taylor expansions. This was already outlined by Poincaré who commented that to find the invariants of a hypersurface one could use "des calculs qui peuvent être longs mais qui restent élémentaires." What is surprising, besides the amount of careful work necessary to carry out this "elementary" computation, is that, as we see in Chapter 4, this computation reveals an underlying geometric structure. Further, this same structure appears in a completely different approach to the problem (Chapters 6 to 9).

The usefulness of weights in this computation is a reflection of the fact that Q is invariant under the map $(z, w) \rightarrow (tz, t^2 w)$ where t is any nonzero real number. The existence of this dilation plays an important role also in other analytic properties of strictly pseudoconvex structures. See, for instance [**BFG**].

Chapter 5

There is a holomorphic version of the Frobenius Theorem which arises from the usual statement and the usual proof simply by assuming that all functions, vector fields, etc. are holomorphic in all their arguments. This version was used in Chapter 1 to establish that all C^ω CR structures are realizable.

There is also a complex version of the Frobenius Theorem. Here the functions, vector fields, etc., are complex-valued but not holomorphic. This is a completely different result and is patterned on the Newlander-Nirenberg Theorem. See [**Ni1**] and [**Ho2**].

Chapter 6

It follows from Cartan's construction that a sufficiently smooth diffeomorphism of real analytic, strictly pseudoconvex CR structures must be real analytic. For let M and M' be real analytic CR structures and let ϕ be a diffeomorphism of M to M'. If ϕ is at least seven times differentiable, then it may be lifted to a map of the geometric bundles B and B'. This lift satisfies a real analytic system of equations of Frobenius form and so is itself real analytic. Thus ϕ is also real analytic. Much stronger results are known. These are related to "reflection principles." See [**BR**] and the references cited therein.

There is a way to directly relate the constructions of Cartan and Moser. Let M be the hypersurface, B the bundle of initial data for the normal form map, and ω_{MC} the Maurer-Cartan connection for SU(2, 1). Fix some point $p \in M$ and some point $b \in B$ in the fiber over p. Consider the mapping to normal form that corresponds to b. This provides an osculation of M to Q and of B to SU(2, 1). The connection ω_{MC} can then be transferred from the fiber of SU(2, 1) over the origin to the fiber of B over p. See [**Ja1**].

Chapter 7

Every C^∞ complex structure is equivalent to a C^ω complex structure. This is not the case for CR structures, as may be shown using the relative invariant r. For it is not difficult to construct some M that has $r = 0$ on an open set without r being identically zero. In any other coordinate system, r would have this same property. But if the structure on M were analytic, then r would be analytic; and this is impossible. (M can be given explicitly in normal form [**Fa**].) However, every realizable CR structure, strictly pseudoconvex or not, is given by the boundary values of analytic functions [**HJ**]. This is a partial converse to the fact that any C^ω structure is realizable.

Chapter 8

The chains on the Reinhardt hypersurfaces R_B have another interesting property. Consider the chains at some point $P_1 = (z_0, w_0)$. For an open set of initial directions the projection $z(t)$ of the chain engulfs the origin. Each such chain contains the point $P_2 = (z_0, e^{2\pi i B} w_0)$ which, for B not an integer, is different from the original point. Thus the two points P_1 and P_2 may be connected by an infinite family of distinct chains. These chains correspond to some subset of the chains on Q that pass through a given point.

Chapter 10

There are now in partial differential equations very general nonsolvability results, which are completely divorced from the several complex variables framework of Lewy's original example. See, for instance, [**Ho1**] and [**NT**]. Also, some very simple examples have been found in two dimensions [**Ni2**]. From the point of view of CR structures, the most important remaining solvability question is that of realizing five-dimensional strictly pseudoconvex manifolds. An interesting recent result here is [**NR**] in which it is shown that, unlike in higher dimensions, there is no "homotopy operator" in dimension five. There is also a new class of nonrealizability results for dimension three [**Ro**].

Finally, time and other constraints precluded the inclusion of several additional topics which might belong in a book such as this. Two which would have been especially appropriate are the Fefferman bundle, from which the chains are realized as the projections of the null geodesics of a Lorentz metric, and Webster's construction of the Cartan curvature, in analogy with the construction of the Weyl conformal curvature from a Riemannian metric. For the first, see [Fef] and [BDS]; for the second [Wel]. Of course, there is also an extensive literature on the analysis of the Lewy operators. A good place to start would be the survey article [BFG] which, in addition, treats some of the topics in this book.

References

[**Ak**] Akahori, T., *A new approach to the local embedding theorem of CR-structures for* $n \geq 4$ (*the local solvability for the operator* $\bar{\partial}_b$ *in the abstract sense*), Memoirs Amer. Math. Soc., Number 366, Amer. Math. Soc., Providence, 1987.

[**AH**] Antreotti, A. and Hill, C. D., *Complex characteristic coordinates and tangential Cauchy-Riemann equations*, Ann. Scuola Norm Sup. Pisa **26** (1972), 299–324.

[**BR**] Baouendi, M. S. and Rothschild, L. P., *A General Reflection Principle in* $\mathbf{C}^2$, to appear.

[**Ba**] Barrett, D., *A remark on the global embedding problem for three-dimensional CR-manifolds*, Proc. Amer. Math. Soc. **102** (1988), 888–892.

[**BFG**] Beals, M., Fefferman, C., and Grossman, R., *Strictly pseudoconvex domains in* $\mathbf{C}^n$, Bull. Amer. Math. Soc. **8** (1983), 125–322.

[**Be**] Bers, L., *Riemann Surfaces*, Courant Institute Lecture Notes, New York University, New York, 1957–58.

[**BdM**] Boutet de Monvel, L., *Intégration des équations de Cauchy-Riemann induites formelles*, Séminaire Goulaouic-Lions-Schwartz, Exposé IX, 1974–1975.

[**Bu**] Burns, D., Jr., "Global Behavior of Some Tangential Cauchy-Riemann Equations", *Partial Differential Equations and Geometry* (Proc. Conf., Park City, Utah, 1977), Dekker, New York, 1979, 51–56.

[**BS**] Burns, D., Jr. and Shnider, S., *Real hypersurfaces in complex manifolds*, in Proc. Sympos. Pure Math—Several Complex Variables, Vol. 30, Pt. 2, Amer. Math. Soc., Providence, 1977, 141–168.

[**BDS**] Burns, D., Jr., Diederich, K. and Shnider, S., *Distinguished curves on pseudo-convex boundaries*, Duke Math. J. **44** (1977), 407–431.

[**BSW**] Burns, D., Jr., Shnider, S., and Wells, R., *On deformations of strictly pseudo-convex domains*, Inventiones Math. **46** (1978), 237–253.

[**Ca1**] Cartan, E., *Leçons sur la géométrie projective complexe*, Gauthier-Villars, Paris, 1928.

[**Ca2**] Cartan, E., *Sur l'équivalence pseudo-conforme des hypersurfaces de l'espace de deux variables complexes*, I, Ann. Mat. **11** (1932) 17–90; II, Ann. Scuola Norm. Sup. Pisa **1** (1932), 333–354 or Oeuvres Complètes, Part II, 1232–1305 and Part III, 1218–1238.

[**Ca3**] Cartan, E., *Les problèmes d'équivalence*, Séminaire de Math., exposé D, 11 janvier, 1937 or Oeuvres Complètes, Part II, 1311–1334.

[**Ca4**] Cartan, E., *Les systèmes différentiels extérieurs et leurs applications géométriques*, Hermann, Paris, 1945.

[**Cat**] Catlin, D., *On the extension of CR structures*, to appear.

[**Ch**] Cheng, J.-H., *Chain preserving diffeomorphisms and CR equivalence*, Proc. Amer. Math. Soc. **103** (1988), 75–80.

[**Che**] Chern, S.-S., *An elementary proof of the existence of isothermal parameters on a surface*, Proc. Amer. Math. Soc. **6** (1955), 771–782.

[**CM**] Chern, S.-S. and Moser, J., *Real hypersurfaces in complex manifolds*, Acta Math. **133** (1974), 219–271.

[**CMa**] Erratum to above, Acta Math. **150** (1983), 297.

[**CH**] Courant, R. and Hilbert, D., *Methods of mathematical physics*, Volume 2, Interscience Publishers, Wiley & Sons, New York, 1962.

[Ea] Eastwood, M., *The Hill-Penrose-Sparling C R-folds*, Twistor Newsletter **18** (1984), 16.

[Fa] Farran, J., *Non-analytic hypersurfaces in* $\mathbf{C}^n$, Math. Ann. **226** (1977), 121–123.

[Fal] Falbel, E., *Nonembeddable C R-manifolds and surface singularities*, thesis, Columbia University, 1990.

[Fe] Federer, H., *Geometric Measure Theory*, Springer, Berlin, 1969.

[Fef] Fefferman, C., *Monge-Ampère equations, the Bergman kernel and geometry of pseudo-convex domains*, Ann. Math. **103** (1976), 395–416; Erratum, **104** (1976), 393–394.

[FK] Folland, G. and Kohn, J., *The Neumann Problem for the Cauchy-Riemann Complex*, Princeton University Press, Princeton, 1972.

[Gr] Greenfield, S., *Cauchy-Riemann equations in several variables*, Ann. Scuola Norm. Sup. Pisa **22** (1968), 275–314.

[GG] Golubitsky, M. and Guillemin, V., *Stable mappings and their singularities*, Springer-Verlag, New York, 1973.

[HJ] Hanges, N. and Jacobowitz, H., *A remark on almost complex structures with boundary*, Amer. J. Math. **111** (1989), 53–64.

[Ho1] Hormander, L., *Differential operators of principal type*, Math. Ann. **140** (1960), 124–146.

[Ho2] Hormander, L., *The Frobenius-Nirenberg theorem*, Arkiv för Math. **5** (1965), 425–432.

[Ho3] Hormander, L., *An Introduction to Complex Analysis in Several Complex Variables*, Van Nostrand, Princeton, 1966.

[Ja1] Jacobowitz, H., *Induced connections on hypersurfaces in* $\mathbf{C}^{n+1}$, Inventiones Math. **43** (1977), 109–123.

[Ja2] Jacobowitz, H., *Chains in C R geometry*, J. Differential Geometry **21** (1985), 163–191.

[Ja3] Jacobowitz, H., *Simple examples of nonrealizable C R hypersurfaces*, Proc. Amer. Math. Soc. **98** (1986), 467–468.

[JT1] Jacobowitz, H. and Treves, F., *Non-realizable C R structures*, Inventiones Math. **66** (1982), 231–249.

[JT2] Jacobowitz, H. and Treves, F., *Aberrant C R Structures*, Hokkaido Math. J. **12** (1983), 276–292.

[Ko] Kobayashi, S., *Transformation Groups in Differential Geometry*, Springer-Verlag, New York, 1972.

[KN] Kobayashi, S. and Nomizu, K., *Foundations of Differential Geometry*, Volumes I and II, Interscience Publishers, 1963 and 1969.

[Koc1] Koch, L., *Chains on C R manifolds and Lorentz geometry*, Trans. Amer. Math. Soc. **307** (1988), 827–841.

[Koc2] Koch, L., *Chains, Null-chains, and C R Geometry*, to appear.

[Kr] Kruzhilin, N., *Local automorphisms and mappings of smooth strictly pseudo-convex hypersurfaces*, Math USSR Izvestiya **26** (1986), 531–552.

[Ku] Kuranishi, M., *Strongly pseudo-convex C R structures over small balls*, Part III, Ann. Math. **116** (1982), 249–330.

[La] Lawson, H. B., *Foliations*, Bull. Amer. Math. Soc. **80** (1974), 364–418.

[LeB] LeBrun, C., *Twistor C R manifolds and three-dimensional conformal geometry*, Trans. Amer. Math. Soc. **284** (1984), 601–616.

[Le1] Lewy, H., *On the local character of the solutions of an atypical linear differential equation in three variables and a related theorem for regular functions of two complex variables*, Ann. Math. **64** (1956), 514–522.

[Le2] Lewy, H., *An example of a smooth linear partial differential equation without solution*, Ann. Math. **66** (1957), 155–158.

[Ma] Malgrange, B., *Sur l'integrabilité des structures presque complexes*, Symposia Math. Vol. 2, INDAM, Rome, 1968, Academic Press, London (1969), 289–296.

[NR] Nagel, A. and Rosay, J. P., *Approximate local solutions of* $\overline{\partial}_b$, *but nonexistence of homotopy formula, for* $(0, 1)$ *forms on hypersurfaces in* $\mathbf{C}^3$, Duke Math. J. **58** (1989), 823–827.

[NV] Neumann, J. von and Veblen, O., *Geometry of Complex Domains*, Institute for Advanced Study, Princeton, reissued 1955.

[NN] Newlander, A. and Nirenberg, L., *Complex coordinates in almost complex manifolds*,

Ann. Math. **64** (1957), 391–404.

[NW] Nijenhuis, A. and Woolf, W., *Some integration problems in almost complex manifolds*, Ann. Math. **77** (1963), 424–483.

[Ni1] Nirenberg, L., *A complex Frobenius theorem*, Seminars on analytic functions I, Princeton University Press, Princeton, N. J., 1957.

[Ni2] Nirenberg, L., *Lectures on Linear Partial Differential Equations*, Amer. Math. Soc., Providence, 1973.

[NT] Nirenberg, L. and Treves, F., *On the local solvability of linear partial differential equations. Part* I: *Necessary conditions*, Comm. Pure Appl. Math. **23** (1970), 1–38.

[Pe] Penrose, R., *Physical space-time and nonrealizable C R structures*, Bull. Amer. Math. Soc. **8** (1983), 427–448.

[Po] Poincaré, H., *Les fonctions analytiques de deux variables et la représentation conforme*, Rend. Circ. Mat. Palermo (1907), 185–220.

[Ro] Rosay, J. P., *New examples of non-locally embeddable C R structures* (*with no non-constant C R distributions*), Ann. Inst. Fourier, Grenoble **39** (1989), 811–823.

[Ros1] Rossi, H., *Attaching analytic spaces to an analytic space along a pseudoconcave boundary*, Proc. Conf. Complex Analysis (Minneapolis, 1964), Springer-Verlag, New York, 1965, 242–256.

[Ros2] Rossi, H., *LeBrun's nonrealizability theorem in higher dimensions*, Duke Math. J. **52** (1985), 457–474.

[Roy] Royden, H., *Real Analysis*, Macmillan, New York, 1963.

[Ru] Rudin, W., *Function theory on the unit ball in* $\mathbf{C}^n$, Springer, Berlin, 1980.

[Sc] Schapira, P., *Solutions hyperfonctions des équations aux dérivées partielles du premier ordre*, Bull. Soc. Math. France **97** (1969), 243–255.

[Sp] Spivak, M., *A Comprehensive Introduction to Differential Geometry*, Volume One, Publish or Perish, Inc., Boston, 1970.

[St] Staudt, K. von, *Beitrage sur Geometrie der Lage* II, Nuremberg, 1858.

[Ta] Taiani, G., *Cauchy-Riemann* (*C R*) *Manifolds*, Pace University, New York, 1989.

[Tan1] Tanaka, N., *On pseudo-conformal geometry of hypersurfaces of the space of n complex variables*, J. Math. Soc. Japan **14** (1962), 397–429.

[Tan2] Tanaka, N., *On non-degenerate real hypersurfaces, graded Lie algebras and Cartan connections*, Japanese J. Math. **2** (1976), 131–190.

[Tr] Treves, F., *On the local solvability of linear partial differential equations*, Bull. Amer. Math. Soc. **76** (1970), 552–571.

[We1] Webster, S., *Pseudo-hermitian structures on a real hypersurface*, J. Differential Geometry **13** (1978), 25–41.

[We2] Webster, S., *A new proof of the Newlander-Nirenberg theorem*, Math. Zeit. **201** (1989), 303–316.

[We3] Webster, S., *On the local solution of the tangential Cauchy-Riemann equations*, Annals Inst. H. Poincaré (Anl) **6** (1989), 167–182, and *On the proof of Kuranishi's Embedding Theorem*, same volume, 183–207.

Subject Index

237